# GEOPHYSICAL MONOGRAPH SERIES

David V. Fitterman, Series Editor
Karl Ellefsen, Volume Editor

## NUMBER 13

# FUNDAMENTALS OF GEOPHYSICAL INTERPRETATION

By Laurence R. Lines
and Rachel T. Newrick

Published by
SOCIETY OF EXPLORATION GEOPHYSICISTS

ISBN 978-0-931830-56-3 (Series)
ISBN 978-1-56080-125-2 (Volume)

Society of Exploration Geophysicists
P.O. Box 702740
Tulsa, OK 74170-2740

Published 2004
Reprinted 2005, 2009
Printed in the United States of America

Library of Congress Cataloging-in-Publication Data

Lines, Laurence R.
    Fundamentals of geophysical interpretation / by Laurence R. Lines and Rachel T. Newrick.
        p. cm. -- (Geophysical monograph series ; no. 13)
    Includes bibliographical references.
    ISBN 1-56080-125-5 (volume) -- ISBN 0-931830-56-7 (series)
    1. Seismic prospecting. 2. Prospecting--Geophysical methods. I. Newrick, Rachel T.
        (Rachel Therese), 1970-
II. Title. III. Series.

TN269.8.L56 2004
622'.1592--dc22
                                            2004056514

# Contents

# About the Authors

**Laurence R. "Larry" Lines** holds B.Sc. and M.Sc. degrees in geophysics from the University of Alberta and a Ph.D. in geophysics from the University of British Columbia. His career includes 17 years with Amoco in Calgary and Tulsa (1976–1993). After a career in industry, Lines held the NSERC/Petro-Canada chair in applied seismology at Memorial University of Newfoundland (1993–1997) and the chair in exploration geophysics at the University of Calgary (1997–2002). Since 2002, he has been head of the Department of Geology and Geophysics at the University of Calgary.

In professional organizations, Lines has served SEG as distinguished lecturer, GEOPHYSICS editor (1997–1999), associate editor, translations editor, publications chairman, and member of the editorial board of THE LEADING EDGE. He has been editor and associate editor of CSEG. Lines and his coauthors have won SEG's award for best paper in GEOPHYSICS twice (1988, 1995) and honorable mention for best paper twice (1986, 1998). Lines holds honorary membership in SEG, CSEG, and the Geophysical Society of Tulsa. He is also a member of APEGGA, EAGE, and AAPG.

**Rachel T. Newrick** earned a B.Sc. in geology (1992) and B.Sc. (Hons.) with first-class honors in geophysics (1993) from Victoria University of Wellington, New Zealand, and a Ph.D. in exploration geophysics from the University of Calgary (2004). During her studies, Newrick worked with BHP Petroleum (Melbourne), Occidental Petroleum (Houston), Veritas (Calgary), and ExxonMobil (Houston).

Newrick won the award for best student paper at GeoCanada 2000. She has served CSEG and CSPG as a convention volunteer. At the University of Calgary, she was president of the Geology and Geophysics Graduate Student Society (2000) and was convenor of the Friday Afternoon Talk Series (2001). She is a member of SEG, CSEG, EAGE, and AAPG. In addition, Newrick is an avid motorcyclist and traveler. She was editor of *WIMA NZ*, the magazine of the Women's International Motorcycling Association, New Zealand branch (1997). She founded a motorcycle club for women in Calgary (1999) and has been a motorcycle instructor since 2000.

# Preface

This book is intended as a text for a geophysical interpretation course for senior undergraduates or first-year graduate students. At the University of Calgary, this course is Geophysics 559, and its enrollment consists of both geology and geophysics students, most of whom will enter the petroleum exploration industry. The course requires some undergraduate knowledge of geophysics.

In teaching this course, we have relied on numerous textbooks, many of which are listed in the references as suggestions for further reading. However, we have found no one book that covers all the recent developments in geophysical interpretation. For this reason, we have developed this textbook, which covers several modern interpretation topics.

This book evolved from the presentation of lectures and laboratory exercises for future exploration geophysicists and geologists at the University of Calgary. The university's Department of Geology and Geophysics is unique in that many of the students are employed in the oil industry while completing their education. Their training requires integration of the disciplines of geology and geophysics in a single interpretation course. The course is presented at a heuristic mathematical level so that the information generally is palatable for earth scientists. Because this text is focused mainly on the challenges of petroleum exploration, most of the content is centered on exploration seismology — the principal geophysical tool of the oil industry.

It would be incorrect to assume that this is the final word on the dynamic science (and art) of interpretation, because the field continues to evolve. Nevertheless, we believe that the concepts discussed in this book will prove useful to geoscientists who enter the fascinating field of geophysical exploration and will serve as a useful starting reference for students in geophysical interpretation.

# Acknowledgments

This book has evolved from the efforts of many colleagues and students, and we acknowledge their valuable contributions. First of all, we thank those who have edited this book. We are especially grateful to Karl Ellefsen of the USGS, who served as SEG volume editor of this book, and to Rowena Mills (SEG manager of GEOPHYSICS, books, and digital publications), who coordinated editing and production. In addition to Karl and Rowena's careful scrutiny of the manuscript, we received essential assistance from Leslie Desreux, Joan Embleton, and Mark Kirtland at the University of Calgary.

Several colleagues and students contributed selected chapters. Chapter 14 is largely the result of an excellent paper in THE LEADING EDGE by Davis Ratcliff, Chester Jacewitz, and Samuel Gray. We thank them for permission to reprint this article. We thank Lines' former colleagues at Amoco, Phil Bording, Adam Gersztenkorn, Gary Ruckgaber, John Scales, Sven Treitel, and Dan Whitmore for collaborative research. Much of Chapters 16 and 17 was paraphrased from our previous papers. We thank Andrew Royle for providing us with a paper for Chapter 19 on AVO. Chapter 24 on cooperative inversion was the result of a paper in GEOPHYSICS by Larry Lines, Alton Schultz, and Sven Treitel. We sincerely thank the Society of Exploration Geophysicists for permission to reprint this paper and many other examples cited in the book.

We thank our families and friends for their patience and support.

Finally, we are grateful to the hundreds of University of Calgary students who have provided valuable feedback on this course material over the years.

Laurence R. Lines
Rachel T. Newrick
Department of Geology and Geophysics
University of Calgary

# Chapter 1

# Introduction

## What is geophysical interpretation?

Geophysical interpretation is a fundamental part of petroleum and mineral exploration. The decision to drill for oil or minerals often depends on our ability to obtain reliable models of the earth by using geophysical data gathered at the earth's surface or in boreholes. Interpretation involves determining the geologic significance of geophysical data and generally integrates all available geologic and geophysical information. It is linked closely to other disciplines such as data acquisition and processing.

Interpretation is a process of estimating an earth model whose response is consistent with all available observations. Examples of geophysical observations might include seismic, gravity, magnetic, electrical, electromagnetic, and borehole data. By this definition, interpretation can be considered a type of geophysical inversion.

Because data sets are always limited in size and scope, several interpretations will be consistent with available data. In other words, our interpretations will show ambiguity, and data fitting alone usually is inadequate. Although interpretation is built on the scientific foundations of geology and geophysics, it is often a fascinating mixture of art and science.

Given the various possible mathematical solutions that describe our data, which interpretation should we use? In such cases, it is helpful to rely on experience in geologic case histories and to have insight into geophysical constraints for our models.

With their experience of related exploration plays as a guide, successful interpreters in exploration generally will choose the most optimistic interpretation that is consistent with available data. Optimism often leads to further data acquisition (for example, shooting more seismic surveys) and more extensive processing, or it will lead to drilling a well. A pessimistic approach often "closes the door" on an exploration play and leads to abandonment of the area as being prospective. The following quotation from Sheriff and Geldart (1995, p. 349) illustrates this point: "Management is usually tolerant of optimistic interpretations that are disproven by subsequent work, but failing to recognize a possibility is an unforgivable sin."

Many factors besides geophysics are critical to commercial production of minerals or oil and gas; for example, the price of the commodity will effectively define whether a discovery is commercially successful. In addition, the expiry date for an exploration lease may dictate the necessity of drilling a well or conducting a seismic program.

Although geophysical interpretation is only one factor in the decision to carry out an exploration program, it is a very important exploration tool. Given that understanding, one might ask the following question: What makes an ideal interpreter?

# The ideal interpreter

In this book, we look at the ideal interpreter mainly from the petroleum exploration perspective, although similar attributes would apply to an interpreter in mining exploration or other applications. The following description of the ideal interpreter is paraphrased from an excellent discussion by Sheriff and Geldart (1995).

The ideal interpreter combines geophysical and geologic information and fully understands the processes involved in generation and transmission of seismic waves, the effects of recording equipment and data processing, and the physical significance of geophysical data. This interpreter's geologic experience would allow for assimilation of massive amounts of data to arrive at the most plausible geologic picture. Such geologic experience requires thorough understanding of rock properties, history of geologic formations, and possible structural and stratigraphic trapping mechanisms.

The *ideal interpreter* probably does not exist. Many interpreters have some of the requisite knowledge and experience in both geology and geophysics, but it would be rare to find all knowledge and experience in a single individual. The best alternative to an ideal interpreter is a team of geologists and geophysicists working in close cooperation. The modern oil company includes a team of petroleum experts working together in synergistic cooperation. For exploration, the team likely would coordinate with the land department of its company. In development and production of a field, the team usually would include reservoir engineers.

Although the terms *teamwork* and *synergies* may be overused in today's exploration environment, they are nonetheless an essential part of most successful oil and gas exploration ventures. On the other hand, successful exploration ventures also rely on the individual's creativity and initiative in interpretation.

## Rock properties and geophysical surveys

Geophysics may be described as "physics of the earth." In the application of geophysics, we make physical measurements that allow us to distinguish between rock formations within the earth. Different geophysical methods allow us to differentiate between rock properties. In petroleum exploration, reflection seismology is the main tool because of its resolving capabilities for most sedimentary rocks. Table 1 briefly summarizes some of the most frequently used geophysical methods for surface recordings and the rock properties they measure.

Although most of the discussions in this course will focus on exploration seismology and most frequently on reflection seismology, the explorationist should be aware of the many geophysical tools that are available. The geophysicist should remember that *seismology is a subset of geophysics*. We hope that the integrated use of geophysical methods will be made clear in the discussions of cooperative inversions of geophysical data.

## Suggested reading

This book is intended as a textbook for a geophysical interpretation course at the senior-undergraduate or first-year graduate level. In teaching this course, we have relied on numerous textbooks, many of which are listed in the references and suggestions for further reading. There are many good exploration geophysics textbooks, such as Sheriff and Geldart (1995). In theoretical seismology, the book by Aki and Richards (2002) is very comprehensive. For seismic processing, Yilmaz (2001) is a popular reference. Formerly, the textbook by Grant and West (1965) provided an excellent reference for interpretation theory in applied geophysics. We also found *An Introduction to Seismic Interpretation,* by McQuillin et al. (1984), to be very worthwhile for all but the most recent developments.

**Table 1.** Frequently used geophysical methods for surface recordings and typical applications.

| Geophysical method | Physical property measured | Typical applications | Comment on applicability |
|---|---|---|---|
| Seismology | Seismic wave velocity, seismic impedance contrast, attenuation, anisotropy | Delineation of stratigraphy and structures in petroleum exploration | Exploration seismology is the most widely used geophysical method in petroleum exploration. |
| Gravity surveys | Rock density contrasts | Reconnaissance of large-scale density anomalies in petroleum and mineral exploration | Gravity surveys are generally less expensive but have less resolving power than seismic exploration. |
| Magnetic surveys | Magnetic susceptibility or the rock's intrinsic magnetization | Reconnaissance of the crustal magnetic properties, especially for determination of basement features | Aeromagnetic surveys are widely used in both petroleum and mining applications for determining large, deep structures. |
| Electrical and electromagnetic surveys | Rock resistivity, capacitance, and inductance properties | Mineral exploration | These methods are used most frequently in mining exploration and well logging (resistivity, SP, and induction logs). |

# References

Aki, K., and P. G. Richards, 2002, Quantitative seismology: University Science Books.

Grant, F., and G. West, 1965, Interpretation theory in applied geophysics: McGraw-Hill Book Co.

McQuillin, R., M. Bacon, and W. Barclay, 1984, An introduction to seismic interpretation: Gulf Publishing Co.

Sheriff, R., and L. Geldart, 1995, Exploration seismology: Cambridge University Press.

Yilmaz, O., 2001, Seismic data analysis: Processing, inversion, and interpretation of seismic data, 2v.: SEG.

# Chapter 2

# Petroleum Reservoirs

A successful explorationist must understand the structure and stratigraphy of exploration targets. We do not generally search for oil or gas directly, but we look for known hydrocarbon traps and assess the probability that they contain hydrocarbon reserves. It is therefore important to understand all the essential ingredients of a petroleum reservoir and to be able to recognize structural and stratigraphic traps, which commonly are searched for in petroleum exploration. Allen and Allen's (1990) detailed discussion of the petroleum play is recommended for further reading.

The essential ingredients of a petroleum system include a source rock, a hydrocarbon migration path, and a reservoir with good porosity, permeability, and a seal. The reservoir and seal form a trap for the hydrocarbons. Oil and gas migrate from a source along porous and permeable pathways until an impermeable barrier seals them in a reservoir. The individual components of the petroleum system are discussed below.

## Source rock

A source rock is rich in organic matter and can produce hydrocarbons if sufficiently heated. The most common organically rich sedimentary rocks for oil and gas generation are shales and carbonates.

Organic matter is deposited in subsiding areas such as lakes, lagoons, deltas, and swamps. The rate of sedimentation must be sufficiently rapid so that the organic matter is partially preserved. Because of poor circulation, the lake/seafloor becomes oxygen deficient or fully anoxic. Oxygen deficiency results in preservation of organic matter.

Oil and gas deposits are formed when rock with a high organic content is cooked (within the hydrocarbon window) over time by the effects of temperature and pressure. The hydrocarbon window is the temperature-depth-maturity conditions that cause source rock to expel hydrocarbons. Figure 1 illustrates the dependence of hydrocarbon generation on the type of organic matter, temperature, and depth of burial.

Type I kerogen, or organic matter, has a marine source and is highly oil prone. Type II kerogen also has a marine source and is oil prone. Type III kerogen is terrestrial in origin and predominantly produces wet gas. Type IV kerogen is generally inert and produces limited dry gas.

As temperature and depth of burial increase to reach the oil window, the source rock starts to mature. In an average sedimentary basin, crude oil is produced when source rock lies within an oil window of 50° to 150°C, with peak generation at about 90°C. Once the oil floor (Figure 1) is reached, oil generation ceases. Gas is generated at higher temperatures and at greater depths than oil.

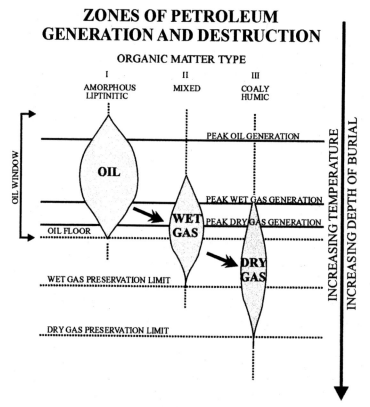

**Figure 1.** Hydrocarbon generation window (modified from Jain and de Figueiredo, 1982).

# Migration path

Hydrocarbon migration is the movement of petroleum from a source rock to a reservoir rock along a path. Migration generally occurs from a structurally low area to a structurally high area because hydrocarbons are buoyant with respect to other pore-fluids rock. Primary migration is the initial expulsion of hydrocarbons from source rock, and secondary migration is the additional movement of hydrocarbons into the reservoir. (The term *migration* also refers to a type of seismic imaging described in later chapters.)

# Trap

A hydrocarbon trap is a body of rock capable of containing hydrocarbons (a reservoir) and sealed by a relatively impermeable barrier (a seal). A reservoir must have sufficient porosity and permeability to allow the flow of hydrocarbons. The seal must be impermeable to the flow of hydrocarbons. Hydrocarbon traps are categorized as either stratigraphic (i.e., pinch-out, unconformity, or reef) or structural (i.e., folded or faulted).

## Porosity

Porosity is the volume of space within a rock that can contain fluids. The amount of porosity dictates the total volume of oil that may be present in a specific reservoir. Rocks with sufficient porosity include sandstones, siltstones, and carbonates. Dolomitized carbonates are preferable because of the large porosity caused by vugs (holes). The porosity in an oil-bearing rock may range from 5% to 30%. It is important to distinguish between effective and noneffective porosity because only effective porosity is interconnected and contributes to fluid flow within a rock (Figure 2). Effective porosity excludes isolated, or noneffective, porosity. Total porosity is the sum of the effective and noneffective porosity.

## Permeability

Permeability is a measurement of the ability of a rock to allow fluid flow. A rock with high permeability will allow a high flow of oil and gas through the rock and generally has high porosity and good connectivity between pores. Conversely, low permeability indicates that the rock may have pores that are small and not well connected. Permeability is measured in millidarcys (md). If permeability is less than 5 md, the reservoir is considered to be tight. Reservoir permeability can be fair (1.0–10 md), good (10–100 md), or very good (100–1000 md) (Levorsen, 1956). Permeability may be a property of the rock matrix, or it can be caused by fracturing. In the latter case, permeability may depend on direction, with good permeability and flow rate aligned with fractures and very little permeability across fractures.

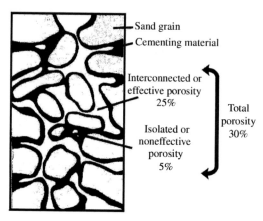

**Figure 2.** Effective, noneffective, and total porosity in a rock matrix (from Jain and de Figueiredo, 1982).

The relationship between porosity and permeability is varied but generally is found to be proportional, as shown by plots of porosity versus permeability for many rock formations (Figure 3a, b).

## Seal

A seal, the final element of the petroleum reservoir, is necessary to prevent further migration of hydrocarbons. A seal is an impermeable rock, often shale or salt, that acts as a barrier to flow. The seal may prevent migration of hydrocarbons by a stratigraphic or a structural configuration.

## Stratigraphic traps

A stratigraphic trap is a hydrocarbon trap caused by lithologic changes rather than structural deformation. Examples of stratigraphic traps are pinch-outs, unconformities, and reefs.

A pinch-out is a stratigraphic trap in which a reservoir unit is surrounded by an impermeable unit or the thinning of a reservoir rock against a seal (Figure 4). Hydrocarbons are trapped within the reservoir. The gas (G), oil (O), and water (W) within the sand body are layered because of their relative buoyancy.

An unconformity trap is any hydrocarbon trap that is bounded by an unconformity (surface of nondeposition). Figure 5 shows a dipping reservoir rock truncated by an impermeable layer. Hydrocarbons migrate along the reservoir layer until they are trapped by the seal. See Chapter 12 for further discussion of unconformities.

Figure 6 shows a schematic representation of hydrocarbons and water within a porous reef and encased by impermeable shales that act as a seal. Identification and characteristics of reefs are discussed further in Chapter

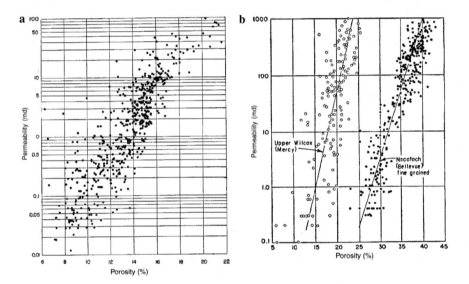

**Figure 3.** Relationship between porosity and permeability for (a) the Bradford sandstone, Pennsylvania, and (b) the Upper Wilcox sandstone, Texas (left), and the Nacatoch sandstone, Louisiana (Levorsen, 1956).

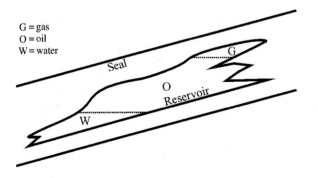

**Figure 4.** A pinch-out is a common type of stratigraphic trap (modified from Jain and de Figueiredo, 1982).

13 because reef exploration makes up a significant portion of hydrocarbon exploration in the Western Canada Sedimentary Basin.

## Structural traps

Structural traps are caused by folding, faulting, or other deformation and include antiformal and various types of faulted traps.

Antiformal traps are formed by strata that are folded convexly upward (Figure 7). The stratigraphic sequence contains an impermeable barrier rock such as shale, preventing further migration of oil trapped within the antiform. An important feature of the trap is the location of the spillpoint, the point at which the trap no longer can hold hydrocarbons and they "spill out."

Faulting causes traps by juxtaposing impermeable strata over porous permeable strata. Sufficient slip has to occur to seal the reservoir fully. Figure 8 contrasts examples of faulted traps with good and bad seals. In the

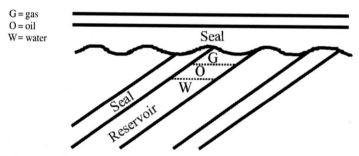

**Figure 5.** An unconformity traps hydrocarbons with an impermeable seal directly above the unconformity.

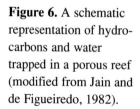

**Figure 6.** A schematic representation of hydrocarbons and water trapped in a porous reef (modified from Jain and de Figueiredo, 1982).

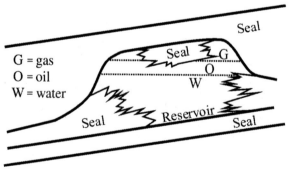

example shown in Figure 8b, the fluids can migrate out of the reservoir. Figure 8a shows a representation of a good seal with respect to juxtaposition of impermeable strata against the reservoir, but the fault gouge may act as a migration pathway, allowing fluids to escape along the fault itself.

## Salt structures

Many trapping mechanisms are associated with salt intrusions, including stratigraphically bounded flank traps, structurally bounded flank traps, fault traps above the salt intrusion, and fault traps along the salt intrusion (Figure 9). Above the salt intrusion, good structural traps are created by extensional folding. Folding associated with the intrusion of salt also may create traps

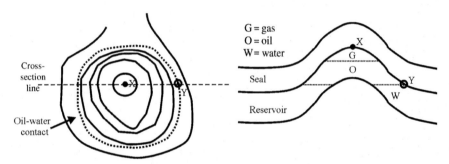

**Figure 7.** Structural trap caused by folding of strata.

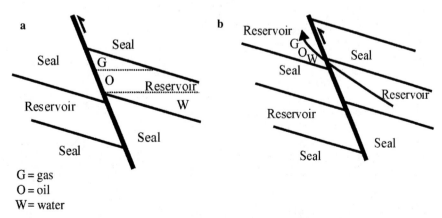

**Figure 8.** An example of a faulted trap (a) with a good seal and (b) without a seal. Note that in (b), gas, oil, and water can migrate out of the faulted region.

**Figure 9.** Many stratigraphic and structural traps are associated with a salt dome (Levorsen, 1956).

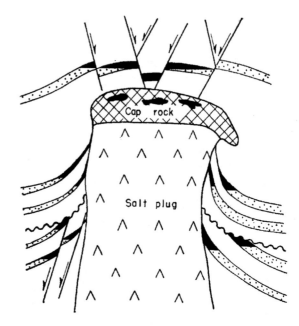

above the salt dome. A seal may be provided either by impermeable strata above the reservoir rock or, in the case of extensional faulting, by the fault gouge itself. Faulting and folding also are observed along the flanks of the salt intrusion.

Additional structural traps are shown in the form of synclinal structures caused by drag folding and truncation against the flanks of the salt dome. Unconformity traps are formed by truncation of the reservoir layer. Impermeable strata overlie the unconformity and prevent further migration of hydrocarbons. There are abundant possibilities for traps related to salt structures. Chapter 14 contains a detailed discussion of the seismic imaging of salt structures.

# References

Allen, P. A., and J. R. Allen, 1990, Basin analysis: Principles and applications: Blackwell Scientific Publications, Inc.

Jain, K. C., and R. J. P. de Figueiredo, 1982, Concepts and techniques in oil and gas exploration: SEG.

Levorsen, A. I., 1956, Geology of petroleum: W. H. Freeman.

# Chapter 3

# Potential Fields

When we think of geophysical interpretation, seismic interpretation often comes to mind immediately. However, many other geophysical methods are used in geophysical interpretation, and we briefly consider two of them here — the gravity and magnetics methods. These are often termed *potential-field* methods because they deal with force fields derivable from potential energy functions. Potential fields have been used in prospecting for minerals in the mining industry and as reconnaissance tools in the petroleum industry.

Nettleton (1971, p. ix) terms the use of gravity and magnetics methods as "the other five percent," because the seismic method accounts for 95% of expenditures in geophysical exploration for petroleum. However, this does not mean that potential-field methods make an inconsequential contribution. Although these methods have less resolving power than seismic prospecting methods, they are generally less expensive. As a rule of thumb, Nettleton gives comparative costs of magnetic, gravity, and seismic data as 1:10:100.

Gravity exploration is sensitive to the anomalous mass of a subsurface body, whereas magnetic surveys are sensitive to magnetic susceptibility of rock bodies. Gravity exploration uses gravimeters, and magnetic surveys use magnetometers. Good descriptions of these instruments are given in Telford et al. (1976). Gravimeters and magnetometers can obtain measurements in the air, on sea, on land, and inside the borehole.

Potential-field methods are useful in many aspects of exploring for petroleum traps, including analysis of source-rock deposition, source maturation, hydrocarbon migration, seals, and timing of emplacement of each component of the petroleum system. These methods generally can be integrated with both borehole logs and seismic data. Johnson (1998) summarizes how gravity and magnetic surveying can assist petroleum exploration (Table 1).

**Table 1.** How gravity and magnetic analyses address petroleum issues (from Johnson, 1998).

| Issue | Gravity and magnetic tasks | Integrated with |
|---|---|---|
| *Source-rock deposition* | | |
| Where were the source rocks deposited? | Depth to magnetic basement | Seismic data |
| How deep are the source rocks? | Regional basin enhancements | Regional geology |
| *Source maturation* | | |
| Where are the "cooking pots" and fetch areas? | Depth to magnetic basement | Seismic data |
| What is the present-day heat influx in the | Isostatic residuals | Well data |
| basin and how much does it vary? | Sediment thickness | Density and velocity data |
| What is the thickness of the crust? | Depth versus density modeling | Heat flow data |
| What is the overburden? | Regional structural modeling | |
| | Curie point (regional heat flow) | |
| | Delineation of volcanics | |
| *Hydrocarbon migration* | | |
| How much relief is there on the basement? | Magnetic inversion | Well and outcrop data |
| What are the "shapes" of the "cooking pots?" | Depth to magnetic basement | Topography |
| Are there major vertical conduits near source areas? | Vertical fault identification | Remote sensing |
| Are there major lineations and how do they relate to | Gradient analysis | Seismic data |
| more recent geologic features? | Regional depocenter and sediment | Sequence stratigraphic analysis |
| | path enhancements | Seismicity |
| *Reservoir prediction* | | |
| Where are the thickest sediments? | Depocenter and sediment path | Seismic data |
| | enhancements | |
| Where is the highest sand probability? | Integrated basin modeling | Lithologic data (outcrop and well) |
| Where was the source of sedimentation? | Density inversion | Sequence stratigraphic analysis |
| What is the influence of tectonics on deposition? | Provenance (magnetic lithology) | Biostratigraphic data |
| | determination | |

**Table 1.** How gravity and magnetic analyses address petroleum issues (from Johnson, 1998) **(cont.)**.

| Issue | Gravity and magnetic tasks | Integrated with |
|---|---|---|
| Have the sediment depocenters shifted over time?<br>What is the compaction history of the sediments?<br>Do the sands have lateral continuity and connectivity? | Sedimentary magnetic analysis<br>Paleomagnetic analysis<br>Integrated velocity analysis<br>(2D and 3D) | |
| *Trap*<br>Where are the major structures?<br>What is the structural grain?<br>Are there faults in the sedimentary sections?<br>Are there lateral porosity changes? | Residuals and enhancements<br>2D/3D structural/stratigraphic modeling<br>Fault identification — gradient analysis<br>Structural inversion<br>Density inversion | Seismic data<br>Outcrop information<br>Topography<br>Remote sensing<br>Seismicity |
| *Vertical seal*<br>Where are salt overhangs?<br>How thick is the tabular salt?<br>How thick are volcanics? | Residuals and enhancements<br>Layer stripping<br>Integrated 2D/3D modeling<br>Sedimentary magnetic analysis | Seismic data<br>Sequence stratigraphic analysis |
| *Timing*<br>What are the ages of the sedimentary features?<br>How do all the petroleum-system elements fit<br>together and what is the timing? | Integrated 2D/3D structural/<br>stratigraphic modeling<br>Layer stripping and enhancements<br>Tectonostratigraphic analysis<br>Paleomagnetic analysis | Density, and velocity data<br>Seismic data<br>Biostratigraphic data<br>Back-stripping<br>Palinspastic reconstructions |

In this chapter, we will briefly describe some aspects of data acquisition, data corrections, simple models of gravity responses, and interpretation of anomaly depths by use of model curves. For more detail on the use of potential fields, several excellent books are available, including Garland (1965), Grant and West (1965), Nettleton (1971), and Telford et al. (1976), as well as many papers on methods and case histories found in journals such as *Geophysics* and *Geophysical Prospecting*.

When the relatively small cost and large benefits of potential-field techniques are considered, we realize that these methods have been underused in geophysical prospecting for new exploration plays. Despite ambiguity in potential-field methods, gravity and magnetics data can impose very definite constraints on geophysical interpretations, as will be shown in Chapter 24.

# Gravity

Gravity surveying relies on the change in gravity caused by a change in mass. Taking two objects with masses, $m_1$ and $m_2$, separated by a distance $r$, the attractive force between them is given by Newton's law of universal gravitation:

$$F_T = -Gm_1m_2 / r^2 . \tag{1}$$

The force is attractive and is a vector directed in the opposite direction to a unit radial vector (hence the minus sign in equation 1). Dividing by mass $m_2$, we obtain the force-per-unit mass that is the gravitational acceleration given by $g = F_T / m_2 = -Gm_1 / r^2$. In the case of the earth, $m_1$ represents the mass of the earth.

When gravity surveying, we use a gravimeter that measures $g_z$, the vertical component of gravitational acceleration. For the case of an object accelerating toward the center of the earth, a typical value would be $g_z = 9.81$ m/s$^2$ or 981 cm/s$^2$ or 981 Gals. (The value of gravitational acceleration typically varies between 978 and 983 Gals over the surface of the earth.) When gravity surveying, we use the units of milligals where 1000 mGals = 1 Gal.

The gravitational acceleration is generally derived from:

$$g_z = G \int \frac{dm}{r^2} \sin\phi , \tag{2}$$

where $\phi$ is the angle of the force vector with the horizontal direction.

Gravity prospecting, therefore, relies on lateral mass differences within the subsurface. And for a specific volume of rock, the contrast in mass is related to the density contrast. Figure 1 illustrates the range of densities for different strata. Considerable overlap exists among the various lithologies, so a priori information is essential for confidence in the final model.

The simplest gravimeter consists of a mass on a spring (Figure 2). The spring is extended, depending on the attraction of surrounding masses. Most springs have a finite length under zero strain. Therefore, when all the strain is removed, the spring is still a certain length. The zero-length spring is of special interest because pretensioning means that the spring has effectively zero length with zero strain. A graph of spring length versus force would pass through the origin. In reality, a spring cannot have "zero length" when external forces are removed; therefore, it must be pretensioned at its minimum length. The zero-length spring used in gravimeters to increase stability and reduce sensitivity to leveling error (Sheriff, 1991) originally was developed by Lucien LaCoste for the LaCoste Romberg style of meter (Figure 3).

## Gravity surveying

Gravity surveying measures the difference in gravity between a known base station and field stations. The procedure is to take a reading at the base station, make field measurements, and then tie back into the base station. By returning to the base station, a calculation for "drift" can be made. Drift results from time changes in the instrument and tidal forces. Using the absolute value for gravity from the base station (given in gravity tables), we

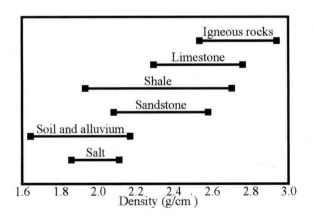

**Figure 1.** Eighty-percent confidence range of small-specimen bulk densities of various kinds of rocks (modified from Grant and West, 1965).

calculate the absolute value of gravity at each field station. It is not necessary to know the actual value of gravity at a station in geophysical exploration — most important are the differences in gravity. We are searching for anomalies caused by density variations in the upper crust.

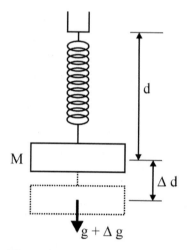

**Figure 2.** A mass on a spring acts as a simple gravimeter.

## Gravity anomalies

A gravity anomaly is the portion of the gravity measurement that deviates from what is expected. There are many types of gravity anomalies, but we are concerned primarily with the Bouguer anomaly. The calculation of Bouguer gravity attempts to remove all gravity effects (such as latitude, elevation, and terrain) that are not caused by local density variations.

Traditionally, the gravity effects caused by latitude, elevation, and terrain have been referred to as "gravity

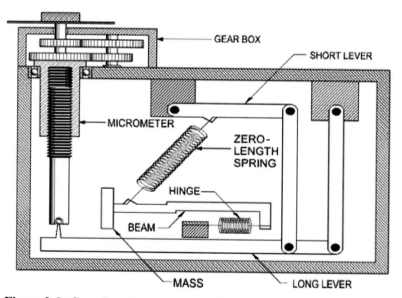

**Figure 3.** LaCoste Romberg gravimeter that uses a zero-length spring (see www.lacosteromberg.com).

corrections" and, for conformity, we continue the use of that term. The gravity corrections are applied to the measured gravity value to obtain a value for the anomaly. It may be helpful to think of the corrections as an estimate of the gravity under specific circumstances, or as "theoretical gravity." For example, the latitude correction is actually an estimate of gravity at a given latitude, the free-air correction and Bouguer correction account for elevation, and the terrain correction includes the effect of undulating terrain. Therefore, the anomaly will simply be the measured gravity minus theoretical gravity.

## Latitude correction

First, gravity is estimated given a specific latitude. The gravity at a point on the earth will vary with latitude because of the varying centrifugal force of the earth's rotation and variations of the earth's radius. Both these effects produce an increase in gravity with increasing latitude. Centrifugal force, which is greatest at the equator, opposes gravitational force, which is therefore a minimum at the equator. In addition, the earth's radius is a maximum at the equator, and gravity is again a minimum (equation 1).

There have been many formulas for latitude correction, but the two generally accepted equations are those that use the 1930 and 1967 reference ellipsoids. Many Bouguer maps have been created using the 1930 reference ellipsoid, but the Geodetic Reference System 1967 (equation 3) is considered to represent a more accurate ellipsoid. More recently, the 1984 reference ellipsoid (equation 4) has been adopted:

$$g(\varphi)\,[1967] = 978031.8\,(1+0.0052789\sin^2\varphi \\ +0.0000235\sin^4\varphi)\,\text{mGal} \tag{3}$$

versus

$$g(\varphi)\,[1984] = 978032.7\,(1+0.0053024\sin^2\varphi \\ +0.0000059\sin^4\varphi)\,\text{mGal}, \tag{4}$$

where $\varphi$ = latitude.

## Free-air correction

Second, we estimate the gravity caused by elevation. As elevation increases, the station is farther from the center of the earth's mass, and gravity decreases. The magnitude of the free-air correction is 0.3086 mGal/m:

$$g(f) = [0.3086h] \text{ mGal},  \tag{5}$$

where $h$ is the elevation of the station in meters.

## Bouguer correction

The free-air correction accounts for the elevation difference but not for the mass between sea level and the station. The mass is accounted for by the Bouguer correction, for which the underlying model is an infinite slab with thickness $h$:

$$g(\beta) = (0.04192\rho h) \text{ mGal},  \tag{6}$$

where $\rho$ is the density contrast between the slab and the material being replaced (i.e., air, water) in g/cm$^3$, and $h$ is the elevation in meters. Once a density is assumed, this correction often is combined with the free-air correction as a single elevation correction.

## Terrain correction

In areas of topographic relief, a terrain correction is made to account for excess mass above the Bouguer slab and a deficit below the slab. Both the excess mass and the deficient mass have the effect of decreasing gravitational force. When we consider a point at the top of the Bouguer slab, we see that the excess mass above the slab has a gravitational effect opposite to the mass of the slab. Therefore, the total gravity effect is decreased. Holes in the slab cause the mass of the slab to be decreased; subsequently, the gravitational effect of the slab decreases. Thus, the combined effect is a decrease in gravitational effect of the slab, and the terrain correction is always opposite in sign to the Bouguer slab correction.

To obtain the terrain correction, the surrounding terrain is divided into sectors and a correction is made, depending on the magnitude of the elevation difference and the proximity to the station. Hammer (1939) describes this method in detail.

## Bouguer anomaly

A gravity anomaly is the difference between the theoretical gravity and measured gravity at a point. Specifically, the Bouguer anomaly is obtained when the theoretical gravity includes the Bouguer slab correction. It is not clear whether a Bouguer anomaly necessarily includes the terrain correction. To specify that a terrain correction has been applied, we use terms such as *terrain-corrected Bouguer anomaly* or *complete Bouguer anomaly*. Reference to a *simple Bouguer correction* suggests that a terrain correction has not been applied. The theoretical gravity is an estimate of gravity at a location where

$$g(\text{theoretical}) = g(\varphi) - g(f) + g(\beta) - g(T), \tag{7}$$

where $g(\varphi)$ is the latitude correction, $g(f)$ is the free-air correction, $g(\beta)$ is the Bouguer slab correction, and $g(T)$ is the terrain correction:

$$g(\text{Bouguer}) = g(\text{measured}) - g(\text{theoretical}). \tag{8}$$

# Gravity modeling

A crucial step in gravity interpretation is to find a model response that matches our observation. That is, we compute forward models to match the gravity anomaly. General gravity-modeling algorithms for arbitrary 3D shapes have been presented by Talwani and Ewing (1960). Computer codes for the Talwani algorithm apply to a wide variety of problems.

Because the gravimeter is sensitive only to lateral changes in mass, there are often many solutions for a given profile. Despite this, gravity anomalies often are modeled as the gravitational response caused by a different geometric shape. Telford et al. (1976) categorized the gravitational response caused by many different geometric shapes, and one of the most interesting of these is the gravitational attraction of a sphere.

## Gravitational attraction of a sphere

The formula for the gravitational attraction of a sphere is both interesting and useful. It is interesting because it represents a good approximation to the gravitational force of the earth, and it is a useful model for many ore bodies.

To derive this formula, we basically use the approach described in Garland (1965). Consider the spherical shell in our Figure 4 (a modified version of the figure taken from Garland's book). To derive the gravitational

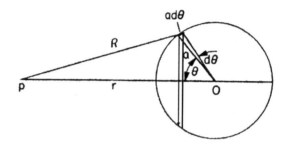

**Figure 4.** Attraction of a spherical shell (modified from Garland, 1965).

formula for the sphere, it is easiest to derive the potential function. If $U$ is the potential, the force is given by its gradient. That is,

$$F = -\nabla U .$$

(9)

We derive the gravitational potential and simply take the derivative of the potential to find the gravitational force.

The gravitational potential, $U$, is given by computing the following formula:

$$U = G\int \frac{dm}{R} .$$

(10)

In this equation, $dm$ represents the unit mass, $G$ is the gravitational constant, and $R$ represents the distance from a point to a unit mass. To compute the gravitational response of a sphere, it is sufficient to compute the response of a spherical shell and consider the sphere as a sum of spherical shells. Figure 4 shows a convenient way to compute this potential for a point that is a distance $r$ from the center of the sphere whose radius is $a$. We consider an elemental ring of the sphere, as shown in Figure 4. All points on the ring are the same distance from observation point O. This distance is given by the trigonometric relationship as

$$R^2 = r^2 + a^2 - 2ar \cos \vartheta .$$

(11)

The radius of the ring is given by $a \sin \vartheta$ where $\vartheta$ is the angle shown in Figure 5. The width of the ring is $ad\vartheta$. If the density of a surface element is $\rho$, then the mass element, $dm$, is given by $\rho(da)(dA)$, where $dA$ is an element of surface area and $da$ is shell thickness. In this case, $dA = 2\pi a \sin \vartheta \, ad\vartheta$.

If we substitute into the previous expression for $U$ in equation 10, we obtain

$$U = Gda\int_0^\pi \frac{2\pi \rho a^2 \sin \vartheta \, d\vartheta}{R} .$$

(12)

From the previous expression for $R$ in equation 11, we see that

$$2RdR = 2ar \sin \vartheta d\vartheta . \tag{13}$$

If we make the substitution of equation 13 into equation 12, we obtain

$$U = Gda \int_{r-a}^{r+a} \frac{2\pi\rho adR}{r} . \tag{14}$$

Therefore, performing the integration gives

$$U = \frac{G\rho 4\pi a^2 da}{r} = \frac{GM}{r} , \tag{15}$$

where $M$ is the mass of the spherical shell. This is a very interesting result, because the potential (and force) of gravitational attraction for a spherical shell is equivalent to placing the entire mass of the sphere at its center. The force of attraction of the spherical shells would be given by the gradient of the potential and is given by

$$F = -\frac{GM}{r^2} . \tag{16}$$

Because of symmetry of the shell and because the force is considered to be the result of a mass at the center of the sphere, we can readily write

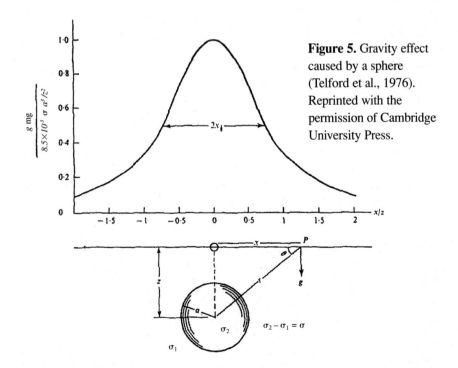

**Figure 5.** Gravity effect caused by a sphere (Telford et al., 1976). Reprinted with the permission of Cambridge University Press.

the formula for a solid sphere. The solid sphere would be a set of shells with the same center. The potential for a solid sphere, W, would be given by integrating equation 15 over a from a = 0 to a = A, where A is the radius of the solid sphere.

$$W = \int_0^A \frac{G\rho 4\pi\, a^2 da}{r} = \frac{G\rho \frac{4}{3}\pi A^3}{r} = \frac{GM}{r}. \tag{17}$$

Because all shells have the equivalent mass at the center, the gravitational attraction of the sphere has the same expression as the shell, where $M$ now represents the spherical mass rather than the mass of a shell.

Let us now examine the gravity response of a buried spherical anomaly for a surface gravimeter, as shown in Figure 5.

The value of $g_z$ is given by multiplying the force-per-unit mass by $\sin\theta$. In terms of Cartesian coordinates $(x,z)$, the distance is given by $r = (x^2 + z^2)^{\frac{1}{2}}$ and $\sin\theta = \frac{z}{r}$. Therefore, the value of the vertical component of gravity is given by

$$g_z = \frac{GMz}{(x^2 + z^2)^{\frac{3}{2}}}. \tag{18}$$

If the density of the sphere is $\rho$ and its radius is $a$,

$$M = \frac{4}{3}\pi\rho a^3 \tag{19}$$

and the gravitational response has a maximum at $x = 0$, in which case,

$$g_{max} = \frac{G\rho \frac{4}{3}\pi a^3}{z^2}. \tag{20}$$

From the response, we can find the $x$ value where $g = g_{max}/2$, and this defines the half-width of the anomaly, $x_{1/2}$. This occurs when $x_{1/2} = 0.77z$. Hence, the depth of the sphere can be estimated from measuring the $x$ position where the gravity response is half of the maximum.

## Other gravity models of interest

Telford et al. (1976) defines several other gravity models of interest. These include the gravity effect of a thin dipping sheet (Figure 6), a vertical cylinder (Figure 7), and a buried slab (Figure 8).

The depths of these and other models can be estimated from interpretation of these model response curves in a manner similar to interpretation of the sphere's depth.

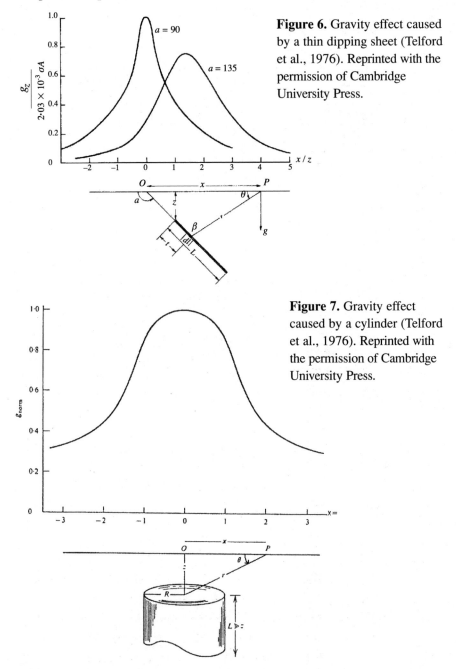

**Figure 6.** Gravity effect caused by a thin dipping sheet (Telford et al., 1976). Reprinted with the permission of Cambridge University Press.

**Figure 7.** Gravity effect caused by a cylinder (Telford et al., 1976). Reprinted with the permission of Cambridge University Press.

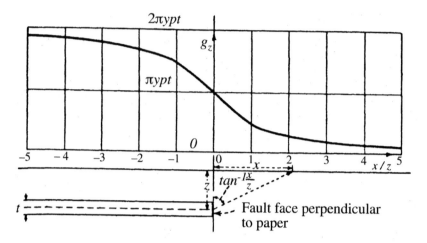

**Figure 8.** Gravity profile over thin faulted slab, downthrown side assumed infinitely deep (Dobrin, 1976).

# Modeling and inversion

As we will discuss in Chapter 16, one of the goals of geophysical inversion (or interpretation) is to produce models whose response matches observations to within the noise level. This is a necessary but insufficient condition to decide whether we have the correct model.

In gravity interpretation, we obtain observations, make the corrections mentioned earlier, filter out certain trends or wavelengths, and then attempt to find models that match the data. As in all geophysical inversions, there will be ambiguities, notably between density and layer depth. Many of these were pointed out by Skeels (1947). Nevertheless, the nonuniqueness of interpretation does not mean that the inversion is not useful, as we shall see in our discussions of cooperative inversion (Chapter 24).

# Magnetics

Our discussions of magnetic prospecting are even briefer than those of gravity prospecting because the aforementioned texts cover this topic.

Modeling the magnetic response caused by a subsurface body is very similar to modeling the gravitational effect caused by a subsurface body. In each case, the response to a geometric shape with specific properties is forward modeled and compared with observed measurements. Magnetic modeling differs from gravity modeling in that vector fields caused by magnetic dipoles are modeled, rather than scalar fields caused by properties of the rock mass.

Poisson related the two fields when the geometry of the causative body is common to the magnetic and gravity sources, showing that magnetic fields are related to gradients of gravitational fields (Grant and West, 1965, p. 213). Therefore, many of the principles of gravity interpretation apply to magnetics interpretation.

One of the main uses of magnetic fields in petroleum exploration is in the definition of basement features in reconnaissance work.

## Aerial surveys

One big advantage of potential-field mapping is its low cost compared to that of seismic data acquisition. This is especially true for aeromagnetic and aerogravity surveys. Aeromagnetic surveys have been in existence for about 50 years. Aerogravity surveys became a popular, if somewhat controversial, topic at the time of a paper by Hammer (1983). Aerial surveys can cover huge areas of inhospitable terrain in short periods of time. High-resolution aeromagnetic (HRAM) surveys have become popular not only for basement mapping but also for mapping of certain sedimentary outcrops. A recent paper by Abaco and Lawton (2003) shows many advantages of HRAM for geophysical prospecting in the Canadian Foothills.

## Conclusions

Although we give only brief discussions here, gravity and magnetics data sets are useful for petroleum exploration. These methods generally are used for reconnaissance to identify anomalies for seismic exploration. Potential fields also can be used effectively for petroleum and mining exploration in areas of poor seismic data. Potential-field data are relatively inexpensive and are worth examining in geophysical exploration.

## References

Abaco, C., and D. C. Lawton, 2003, Aeromagnetic anomalies from the South-Central Alberta foothills: Canadian Society of Exploration Geophysicists Recorder, **28**, no. 1, 26–32.

Dobrin, M., 1976, Introduction to geophysical prospecting, 3rd ed.: McGraw-Hill.

Garland, G. D., 1965, The earth's shape and gravity: Pergamon Press.

Grant, F., and G. West, 1965, Interpretation theory in applied geophysics: McGraw-Hill.

Hammer, S., 1939, Terrain corrections for gravimeter stations: Geophysics, **4**, 184–194.

Hammer, S., 1983, Airborne gravity is here: Geophysics, **48**, 213–222.

Johnson, E. A., 1998, Gravity and magnetic analyses can address various petroleum issues: The Leading Edge, **17**, 98.

Nettleton, L. L., 1971, Elementary gravity and magnetics for geologists and seismologists: SEG.

Sheriff, R., 1991, Encyclopedic dictionary of exploration geophysics, 3rd ed.: SEG.

Skeels, D. C., 1957, Ambiguity in gravity interpretation: Geophysics, **12**, 43–56.

Talwani, M., and M. Ewing, 1960, Rapid computation of gravitational attraction of three-dimensional bodies of arbitrary shape: Geophysics, **25**, 203–225.

Telford, W. M., L. P. Geldart, R. E. Sheriff, and D. A. Keys, 1976, Applied geophysics: Cambridge University Press.

# Chapter 4

# Refraction Seismology

In modern times, refraction seismology has been used to examine the near surface in petroleum exploration and engineering applications. In refraction seismology, the relevant seismic arrivals are the direct wave and the "head wave" (critically refracted arrival). For ease of interpretation, the head wave is identified most easily when it is the first event on a seismogram.

An understanding of the direct arrival and the head wave can be obtained by examining Figure 1. Detailed explanations of refraction seismology are included in many texts, including Grant and West (1965) and Kearey and Brooks (1991).

In homogeneous media such as the one in Figure 1, we can think of seismic energy emanating from a point source as a spherical wavefront. In the 2D model of the figure, these become circles. For P-waves, rays constructed perpendicular to wavefronts will describe the direction of wave motion. Consider the wave motion in the two-layer model of Figure 1 for the case in which the velocity of the second layer is greater than that of the first layer. Let the top layer have a thickness $h$ and a velocity $v_1$, and let the velocity of the second layer be $v_2$.

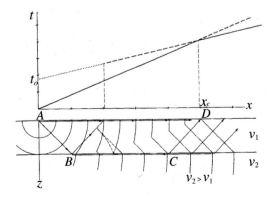

**Figure 1.** Diagram (modified from Grant and West, 1965) showing wavefronts and rays for the direct arrivals and head waves that transmit from a source to a set of geophones.

Recall that the refraction of raypaths at a boundary between two layers is governed by Snell's law:

$$\frac{\sin\theta_1}{v_1} = \frac{\sin\theta_2}{v_2}.$$

(1)

Here, $\theta_1$ and $\theta_2$ are angles between the ray direction and a normal vector to the surface. From equation 1, note that larger velocities imply larger values of $\theta$. Consider the case in the figure in which $v_2 > v_1$ so that the angle of the refracted ray in the second layer is always greater than the angle of incidence. As the incidence angle increases, $\sin\theta_2$ eventually will become 1, and the ray will travel parallel to the layer so as to be refracted critically. The angle at which this occurs is termed the critical angle and, from equation 1, is given by

$$\theta_1 = \sin^{-1}\left(\frac{v_1}{v_2}\right).$$

(2)

Note that the wavefront emanating from the second layer is a plane wavefront. As shown by Cerveny and Ravindra (1971), this can be explained in terms of Huygen's wavefronts. The plane wavefront sent upward from the boundary is the envelope of the Huygen's wavefronts.

Write down the traveltimes of the first arrivals for seismic energy traveling from a source to a set of receivers at the surface. Consider the traveltime, $t$, for an offset value of $x$. The direct arrival will be the first arrival for the near geophone. It travels only in the first layer, and its traveltime is given by

$$t = \frac{x}{v_1}.$$

(3)

The head wave will travel farther but faster in the second medium. Its time of travel will be given by considering the raypath in three segments. The endpoints of the segments are labeled A, B, C, and D in Figure 1. For the head wave,

$$t = t_{AB} + t_{BC} + t_{CD}.$$

(4)

The first and third terms are equal because of symmetry. If we combine these two terms, we obtain a two-term expression for the traveltime,

$$t = \frac{2h}{v_1\cos\theta_1} + \frac{1}{v_2}(x - 2h\tan\theta).$$

(5)

If we rearrange terms and use trigonometric identities we get

$$t = \frac{x}{v_2} + \frac{2h\cos\theta_1}{v_1}. \tag{6}$$

If we then use Snell's law to replace $\cos\theta_1$ in terms of velocities, we obtain

$$t = \frac{x}{v_2} + \frac{2h\sqrt{v_2^2 - v_1^2}}{v_1 v_2}. \tag{7}$$

Equation 3 gives the traveltime for the direct wave. Equation 7 gives the traveltime for the head wave in terms of layer thickness and velocities. For small offsets, the first arrival will be caused by the direct wave. For larger offsets, the first arrival will be caused by the head wave.

Figure 2 gives an example from McAulay (1986) that shows first arrivals on a seismic shot record. The direct arrivals and the head-wave traveltimes are marked by two lines on this seismogram. The picked times would be plotted in a manner similar to those in Figure 1. To find velocities and layer thicknesses, one could then measure the slopes, the intercept time for the head wave, and/or the crossover distance (where the lines cross).

There are several ways to use these equations to determine velocities and thicknesses of layers. Using the direct-arrival times described by equation 3, the slope of the traveltimes for the direct wave equals the reciprocal of $v_1$, the velocity for the first layer. Similarly, the slope of the head-wave

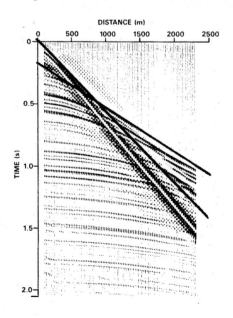

**Figure 2.** Seismogram (from McAulay, 1986) clearly showing the direct and head-wave arrivals, marked by dark lines.

arrivals described by equation 7 produces the reciprocal of $v_2$, the velocity of the second layer.

The thickness of the top layer, $h$, can be derived by two methods. One of these would use equation 7 and find the time axis intercept at $x = 0$. If this value of the intercept is estimated, then $h$ can be expressed in terms of the estimated velocity values. A second approach would be to determine the offset value, where the direct-arrival time equals the head-wave time. This point is defined by the change in slope of the arrival times and is known as the crossover distance, $x_c$.

At this point, we use the definition of crossover distance and equate equation 3 to equation 7,

$$\frac{x_c}{v_1} = \frac{x_c}{v_2} + \frac{2h}{v_1 v_2}\sqrt{v_2^2 - v_1^2}\,. \tag{8}$$

If we solve this expression for thickness $h$ in terms of the layer velocities and the crossover distance, we obtain

$$h = \frac{x_c}{2}\sqrt{\frac{v_2 - v_1}{v_2 + v_1}}\,. \tag{9}$$

Therefore, for a horizontally layered earth model, we can determine the velocities and thicknesses of the layers by measuring the slopes on the $x$-$t$ graph. If the layers are dipping, we must do a forward and reverse profile and measure the slopes to find the dip of the boundary between layers 1 and 2.

In exploration seismology, the refraction method is used primarily for finding the thickness and velocities of the upper layers. This is very useful for determining the statics corrections that could affect arrival times for deeper reflections.

# References

Cerveny, V., and R. Ravindra, 1971, Theory of seismic head waves: University of Toronto Press.

Grant, F., and G. West, 1965, Interpretation theory in applied geophysics: McGraw-Hill.

Kearey, P., and M. Brooks, 1991, An introduction to geophysical exploration: Blackwell Science.

McAulay, A. D., 1986, Plane-layer prestack inversion in the presence of surface reverberation: Geophysics, **51**, 1789–1800.

# Chapter 5

# Reflection Seismology Concepts

## Overview

In the oil industry, geophysical interpretation is concerned mainly with one method — surface reflection seismology. Although other seismic techniques, such as refraction seismology, borehole seismology (cross-borehole and vertical seismic profiling [VSP]), and teleseismic methods are sometimes used, reflection seismology remains the workhorse of exploration geophysics.

Many good textbooks describe theoretical seismology (Aki and Richards, 2002), exploration seismology (Sheriff and Geldart, 1995), and seismic data processing (Yilmaz, 2001). Our purpose is not to duplicate the excellent information on exploration seismology in those books but to explain some basic principles of reflection seismology that are necessary for seismic interpretation. These principles involve reflection amplitudes and traveltimes. We leave the details to those other textbooks.

## Reflection coefficients

Reflections from within the earth arise because of changes in seismic velocity or rock density or both. The simplest model involves a series of rock layers. Consider a sound wave sent downward into the earth by an explosion or vibratory disturbance. In exploration seismology, these waves usually are generated by dynamite or vibroseis sources.

If these waves encounter a discontinuity in acoustic impedance (the product of rock density and seismic velocity), this results in a reflected wave. The ratio of the reflection amplitude to the incident-wave amplitude is given by the reflection coefficient. By solving the boundary conditions for layer discontinuities for an acoustic wave at vertical incidence (as described in Robinson and Treitel, 1980), we can derive the reflection coefficient as being

$$r = \frac{\rho_2 v_2 - \rho_1 v_1}{\rho_2 v_2 + \rho_1 v_1}, \tag{1}$$

where $\rho_1$ and $v_1$ represent the density and velocity of the first layer in which the wave travels and $\rho_2$ and $v_2$ represent the density and seismic velocity of the second layer (see Figure 1).

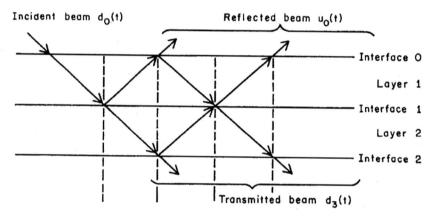

**Figure 1.** Reflected and transmitted waves from an incident wave (from Robinson, 1967).

# Convolutional model

The convolutional model provides a description of the primary reflections in a seismic trace. It represents a way to do the bookkeeping on a set of reflected seismic pulses in a seismogram. For the time being, we will ignore multiple reflections.

We know that our seismogram is not a series of knife-sharp arrivals. There is a finite response time for any seismic disturbance, because the earth does not respond instantaneously to an impulsive disturbance. The earth's displacement includes compression, followed by a rebound or rarefaction. Most seismometers (geophones) respond to the particle velocity of the ground around the geophone. The source pulse, or seismic source wavelet, is the time-varying displacement applied to the ground around the source.

The recorded seismic trace can be viewed as a series of reflected source wavelets that reach a receiver at different delay times. In other words, the recorded trace is a sum of echoes. Let us view this in physical terms and show that the mathematical description is a convolution.

For the sake of simplicity, consider an earth with three layers of equal traveltimes, with two-way traveltimes equal to the sample interval. That is, we consider two reflecting boundaries in which the reflection time series is $(r_0, r_1)$. Consider a three-term source-wavelet sequence whose amplitudes are described by a time series $(w_0, w_1, w_2)$ that is sent down into the earth. Let us consider the seismic trace, $y(t)$, a time series of primary reflections recorded by a geophone just above the top layer.

The first sample is given by the start of the wavelet reflected off the top layer. The first sample of the trace is given by

$$y_0 = w_0 r_0. \tag{2}$$

The second time sample is given by the sum of two arrivals, the first wavelet value bouncing off the second reflector and the second wavelet value bouncing off the top reflector:

$$y_1 = w_0 r_1 + w_1 r_0. \tag{3}$$

The third and fourth samples of $y(t)$ are given by $y_2 = w_1 r_1 + w_2 r_0$ and $y_3 = w_2 r_1$, respectively.

We note that the seismic trace consists of sums of products. Note that the sums of the subscripts of terms in the series have the same value. This sum is equal to the index of the trace-sample index. We generally can write the time series for the seismic trace in terms of the source wavelet and the reflectivity time series by the following:

$$y_t = \sum w_\tau r_{t-\tau}. \tag{4}$$

This equation expresses the discrete convolution of a time series, $w_t$, with a time series, $r_t$, and it is expressed symbolically as

$$y_t = w_t * r_t. \tag{5}$$

The process of convolution can be described in such words as *flip, shift, multiply,* and *add.* That is, the order of one of the time sequences is flipped or reversed in time. It is then shifted by some amount; $t$, the terms in the time series are multiplied; and the products are added together.

We can compute a convolution by applying the above formula, or we can achieve the same result by use of polynomials known as *z*-transforms.

The $z$-transform of a time series is simple to compute. For example, if we have a time series for the wavelet given by ($w_0$, $w_1$, $w_2$), we write down its $z$-transform by considering a polynomial whose coefficients of powers of $z$ are given by the time-series values. In other words, the $z$-transform for the wavelet time series is given by $W(z) = w_0 + w_1z + w_2z^2$. Likewise, the $z$-transform for the reflectivity series in our example is given by $R(z) = r_0 + r_1z$. We note that the $z$-transform of the seismic trace is given by

$$Y(z) = W(z)R(z). \tag{6}$$

We can verify that this is the case by multiplying these polynomials and comparing the coefficients of the resulting $z$-transform to those derived in the previous paragraph.

If we set $z = e^{-i\omega}$ in the previous expression, the $z$-transforms become discrete Fourier transforms. Therefore, we see that convolution in the time domain can be achieved through multiplication of Fourier transforms in the frequency domain.

Before we leave convolutions, let us briefly revisit the seismic trace model. In our discussion, we considered only primary reflections. However, we know that multiple reflections exist in seismograms and can be quite troublesome. When can multiples be safely ignored, and when do we need to account for them or take steps to suppress them?

Because multiples involve two or more reflections off layer discontinuities, their amplitudes are related to products of reflection coefficients — one reflection coefficient for every bounce off a reflector. For typical values of rock densities and seismic velocities, most seismic-reflection coefficients are not greater than 0.1, which means that multiples typically would have products of reflection coefficients that are 0.01 or less. Nevertheless, multiples can be troublesome, especially in the marine case in which the air-water interface has a reflection coefficient of nearly 1.0 and the water-rock interface has a reflection coefficient that is usually greater than 0.1. We can model the response of a layered medium for primaries and multiples by using the techniques outlined by Robinson and Treitel (1980).

Seismic processing can allow us to attenuate multiples, in that multiples generally have a predictable nature and generally have a slower apparent velocity than primaries arriving from deeper reflectors. This will be discussed further in later chapters, where we show that predictive deconvolution and normal moveout followed by stacking can prove useful in suppression of multiples.

# Traveltime, velocity, and normal moveout

It has been said that the most useful equation in seismology is $d = vt$, or distance = velocity × time. In exploration seismology problems, we want to determine the distance of drilling targets by using seismograms whose responses are measured in time. Hence, velocity of seismic-wave propagation provides the essential link between the desired objective and the time measurements of seismic energy.

It can be argued that seismic velocity estimation is the most important problem in exploration seismology because it impacts both processing and interpretation of seismic data. In reflection seismology, the velocity usually is estimated by a technique known as normal moveout (NMO).

To understand NMO requires a simple knowledge of Pythagoras' theorem. In Figure 2, we consider a primary reflection that travels from a seismic source down to a flat seismic reflector and back to a receiver at the surface. Let the distance from the source to receiver be denoted by $X$, the midpoint by M, the depth to the reflector by $Z$, and the distance from the source to the reflecting point by $SR$. For a flat reflector, the downward path is the mirror image of the upward path. By Pythagoras' theorem, we note in the raypath figure that we have a right-angle triangle whose sides are related by

$$|SR|^2 = Z^2 + \left(\frac{X}{2}\right)^2.$$ (6)

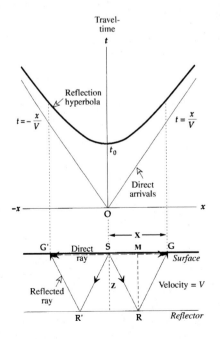

**Figure 2.** Hyperbolic traveltime versus distance curve (top) is shown for reflections from a horizontal boundary (modified from Lowrie, 1997).

On a seismogram, we measure the two-way reflection time, $T$, for a given source-receiver distance, $X$. For the case of zero offset where $X = 0$, we measure the two-way normal-incidence reflection time of $T_0$. The distances in the Pythagorean relationship can be rewritten in terms of traveltimes using $SR = vT / 2$ and

$$(vT / 2)^2 = (vT_0 / 2)^2 + (X / 2)^2 . \tag{7}$$

This produces the well-known hyperbolic relationship among time, distance, and velocity for a primary reflection. In other words, for flat reflectors, reflections will have the traveltime trajectory whose shape is a hyperbola governed by the following equation:

$$T^2 = T_0^2 + X^2 / v^2 . \tag{8}$$

By inspection of the above equation, it is evident that the difference between the traveltime for nonzero offset, $T$, and the normal incidence time, $T_0$, is governed by the offset distance and the velocity. This difference is termed the normal moveout and is given by $\Delta T = T - T_0$.

Much of classical velocity estimation is given by NMO analysis. To understand this more completely, let us examine the mathematical expression for NMO:

$$\Delta T = T_0 (1 + \frac{X^2}{v^2 T_0^2})^{1/2} - T_0 . \tag{9}$$

The values of the NMO, $\Delta t$, as a function of offset and normal incidence time, can be used to estimate the velocity, $v$. There is a convenient relationship between velocity and NMO for values of offset that are small relative to depth. For such cases, we can use the binomial approximation that

$$(1 + \frac{X^2}{v^2 T_0^2})^{1/2} \cong 1 + \frac{X^2}{2v^2 T_0^2} . \tag{10}$$

Making this substitution into the previous expression for $\Delta T$, we obtain

$$\Delta T \cong \frac{X^2}{2v^2 T_0} . \tag{11}$$

For such cases, the velocity can be determined directly from the NMO, normal-incidence reflection time, and offset from the expression

$$v = \frac{X}{\sqrt{2 \Delta T T_0}} . \tag{12}$$

At first glance, it might seem that a single flat-layer model would be too simplistic for use in a real-world situation of many layers. However, an analysis of the multilayered case by Taner and Koehler (1969) showed that the traveltime equation for several layers is in the same form as a single layer, except that velocity, $v$, is replaced by the root-mean-square velocity, $v_{rms}$. That is,

$$T^2 = T_0^2 + \frac{X^2}{v_{rms}^2} .$$ (13)

In addition, if we determine the root-mean-square velocity for two successive reflectors, we can use Dix's equation to determine the interval velocity of the layer between these reflectors. The reader is referred to the classic paper by Dix (1955).

## Seismic data acquisition

In initial presentations of this interpretation course, the authors developed a short chapter on data acquisition. During that year, however, an SEG monograph published by Evans (1997) covered topics in seismic acquisition in more detail.

## Seismic data processing

Seismic processing is covered in detail by Yilmaz (2001). After the assignment of geometry and demultiplexing of field tapes, we proceed with statics corrections, velocity analysis, and NMO corrections. The summation, or "stacking," of common reflection-point data is a very important step for the reduction of noise. The steps of deconvolution and migration improve seismic resolution in the temporal and spatial sense. These steps are covered in other sections of these course notes.

## References

Aki, K., and P. G. Richards, Quantitative seismology: Theory and methods: University Science Books.

Dix, H., 1955, Seismic velocities from surface measurements: Geophysics, **20**, 68–86.

Evans, B. J., 1997, A handbook for seismic data acquisition in exploration: SEG.

Lowrie, W., 1997, Fundamentals of geophysics: Cambridge University Press.

Robinson, E. A., 1967, Multichannel time series analysis with digital computer programs: Holden-Day.

Robinson, E. A., and S. Treitel, 1980, Geophysical signal analysis: Prentice-Hall. Reprint, 2000, SEG.

Sheriff, R., and L. Geldart., 1995, Exploration seismology, 2nd ed.: Cambridge University Press.

Taner, T., and F. Koehler, 1969, Velocity spectra — Digital computer derivation and applications of velocity functions: Geophysics, **35**, 551–573.

Yilmaz, O., 2001, Seismic data analysis: Processing, inversion, and interpretation of seismic data, 2 v.: SEG.

# Chapter 6

# Seismic Resolution

## Overview

The seismic method is limited in its ability to resolve small anomalies. The definition of "small" is governed generally by the seismic wavelength. Wavelength is given by the ratio of phase velocity to temporal frequency (i.e., $\lambda = v/f$). Because there is little we can do about a rock's seismic velocity, we can change the seismic wavelength only by changing the temporal frequency. We shall see that reducing the wavelength by increasing the frequency will help improve both vertical and horizontal resolution.

## Vertical resolution — How thin is a thin bed?

The problem of vertical resolution was posed in a classic paper entitled "How thin is a thin bed?" by Widess (1973). It is crucial in any exploration play to know the limits of resolution. For example, can we hope to define a sandstone layer of 10-m thickness if its seismic velocity is 3000 m/s? The answer will depend on the dominant seismic frequency in our data.

The Widess analysis of this question was obtained by a simple model in which a layer of velocity, $v_2$, is contained in a medium whose velocity is $v_1$, as shown in Figure 1. If the reflection from the top of the layer is given by a reflection coefficient $R_1$, then the reflection from the bottom of the layer is given by $-R_1$. For the moment, let us ignore transmission effects and subsequent multiples. For small reflection coefficients, these will be second-order effects. If our wavelet were a delta function, the seismogram would contain an arrival of amplitude $R_1$ followed by a time-delayed amplitude of $-R_1$. The time delay would be given by the two-way vertical traveltime, which would be twice the layer thickness divided by the seismic velocity.

The resolution question becomes interesting and more realistic when we deal with wavelets of a dominant frequency (and wavelength), because this allows us to relate resolution to frequency. Widess did this by considering the situation shown in Figure 1, in which a bed of thickness $b$ is of velocity 20,000 ft/s (6000 m/s) and is between layers with velocity of 10,000 ft/s (3000 m/s). The reflection coefficients for the top and bottom of the layer are 0.333 and –0.333, respectively. If we ignore the transmission losses caused by reflection (about 11%), the reflections from the top and bottom of the layer would be equal in magnitude and opposite in sign. Widess considered the reflected wavelets to be cosine waves of a particular frequency. (This is also an approximation, because physical wavelets would be causal, would be finite in duration, and would contain a wider band of frequencies.)

**Figure 1.**
Resolution of a
thin bed, as
illustrated by
Widess (1973).

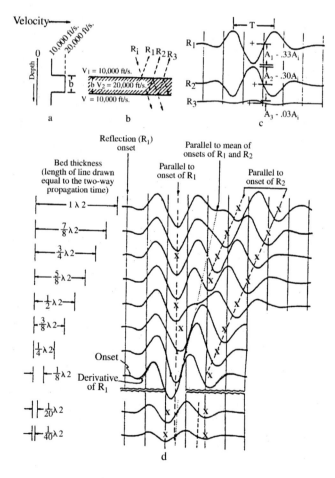

If we consider time zero to be, for a reflection from the middle of the bed, of thickness $b$ and velocity $v$, the reflection from the top of the bed would be given approximately by

$$R_1 \cong A\cos[2\pi f(t+b/v)], \tag{1}$$

and the reflection from the bottom of the bed would be delayed by a two-way time of $2b/v$ with an amplitude of opposite sign. Its reflection response is given by

$$R_2 \cong -A\cos[2\pi f(t-b/v)]. \tag{2}$$

Ignoring higher-order multiples, the sum of the reflections would be given by

$$\begin{aligned} R_d &= R_1 + R_2 \\ &\cong A\{\cos[2\pi f\ (t+b/v)]-\cos[2\pi f(t-b/v)]\} \end{aligned} \tag{3}$$

If we use the formula for the difference of cosines, equation 3 can be simplified to give

$$R_d \cong -2A\sin(2\pi ft)\sin(2\pi fb/v). \tag{4}$$

Now, if our bed thickness is quite small relative to wavelength (i.e., if $b \ll v/f$), we can set $\sin(2\pi fb/v \cong 2\pi fb/v$. Then

$$R_d \cong -\frac{4\pi fb}{v}\sin(2\pi ft). \tag{5}$$

As shown by equation (5), the amplitude of the sum of the reflected arrivals is proportional to the bed thickness ($b$) and inversely proportional to the wavelength ($\lambda = v/f$). In other words, the reflection amplitude question relates to the ratio of bed thickness to the seismic wavelength.

How thin can bed thickness be (relative to wavelength) and still be recognizable as two distinct events? The answer to this question also is illustrated by Figure 1. We note that if the bed is the thickness of one wavelength, the events are easily separable. Reflected events do not overlap until thickness of the bed is about half the wavelength, because the distance for two-way travel is equal to that of the wavelength. As the bed gets thinner and thinner, equation 5 applies. We note that the reflection response acts as a differentiator (turning cosines into sines with opposite polarity).

Based on the seismograms shown in Figure 1, Widess claimed that thickness of a bed would be barely detectable if the bed had a thickness of $\lambda/8$. This separation is the point at which the summation of wavelets is essentially a derivative of the incident wavelets. This resolution limit of $\lambda/8$ often is considered a limit that would be valid under ideal circumstances.

Thus, the Widess paper is important, but it is certainly not the final word on resolution.

A paper that extended Widess' analysis was presented by Kallweit and Wood (1982). Widess had analyzed the case for reflections of opposite polarity. Of course, we could have situations of velocity increase of the same sign at the top and bottom of the reflecting bed, as Kallweit and Wood pointed out. In such cases, we would deal with reflections of the same polarity. This could cause a different type of tuning with arrivals of like sign, as shown by Figure 2. To determine the limits of resolution in this case, Kallweit and Wood examined Rayleigh's resolution criterion from the theory of optics.

If the high-frequency models of geometric optics were perfect, optical instruments could completely separate beams of light. However, in optics, as in seismology, geometric optics is but an approximation to wave theory. We deal not with spike (delta functions) responses but with finite wavelets, and our ability to separate interfering waveforms depends on their frequency.

Figure 2 shows interfering wavelets for the case of reflection coefficients with the same sign. In the resolved case, where reflection coefficients have sufficient separation, we see two separated wavelets that are easily identifiable. As we decrease the separation of spikes (or decrease the thickness of a reflecting bed), we note that the wavelets merge into a single spike. Just as in the Widess example, the question of resolution boils down to the minimum separation of wavelets that is required to identify separate

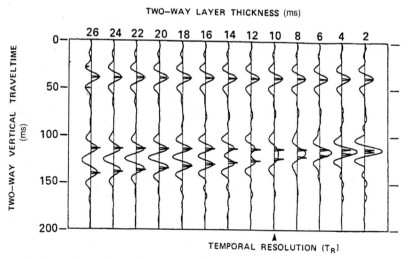

**Figure 2.** Resolution of arrivals as a function of reflector separation (from Kallweit and Wood, 1982).

reflections. Denote the wavelet breadth by $B$ (distance between successive troughs or peaks). $B$ is also equal to the wavelength if our wavelet contains a single frequency. The optical limit established by Rayleigh (and derived in Jenkins and White, 1957, p. 300) is $B/2$, which is the peak-to-trough separation of the diffraction pattern for an optical instrument.

Figure 3 shows the appearance of the overlapping wavelets with Rayleigh's criterion. Ricker (1953) noticed the merging of overlapping wavelets for reflection coefficients of like sign and chose a slightly different and more optimistic criterion than that chosen by Rayleigh. Ricker chose the resolution point to be the point where the composite wavelet had a "flat spot," or zero curvature, as shown in Figure 3. For Ricker wavelets, this flat spot occurs when the time separation equals $1/(2.3F)$, where $F$ is the predominant frequency of the Ricker wavelet. (The derivation is given in Kallweit and Wood, 1982.) In terms of wavelength, the resolution criteria are $\lambda/4.6$ for the Ricker criterion and $\lambda/4$ for the Rayleigh criterion.

After discussing the Widess, Rayleigh, and Ricker criteria for resolution, the interpreter is still left with the questions: Which criterion should I use? How thin is a thin bed in my area? The choice of any of the criteria depends on knowledge of the frequency content in the data. In deciding on the limits of resolution, one would usually take the following approach of computing a synthetic seismogram. Find a sonic and density log from a well in the area of investigation and convert the logs to acoustic impedance as a function of time. Compute the reflectivity for these logs by differencing the impedance log. This will produce the ideal reflectivity up to the Nyquist fre-

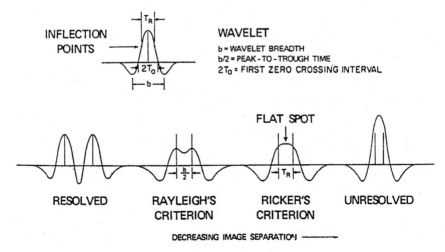

**Figure 3.** Resolution criteria, as illustrated by Kallweit and Wood (1982).

quency of the sampled log. Finally, convolve the reflectivity sequence with a zero-phase wavelet whose frequency content is similar to the frequency content of seismic data in the area. Then, by inspection of the reflected arrivals for thin beds in the synthetic seismogram, we can begin to know whether we can hope to resolve a thin bed.

# Improving temporal resolution through deconvolution

After modeling the reflection response of a thin bed, an interpreter should know whether the seismic data will allow for resolution of a thin bed. If the thin bed can be resolved, the interpreter can proceed with interpretation. If the thin bed is too thin for resolution, then the interpreter must think of possible remedies (other than giving up hope).

We note that all resolution criteria are dependent on the seismic wavelength, which we need to reduce if we hope to improve resolution. It is sobering to note that a typical seismic wavelength in exploration seismology may be the length of a football field (about 100 m). Wavelength is given by velocity/frequency. There is not much we can do to change seismic velocities. Therefore, to reduce wavelengths and improve resolution, we need to increase the high-frequency content in our data. This generally is achieved by seismic deconvolution.

The seismic trace is modeled by the convolution of a wavelet with the reflectivity response. The seismic wavelet is a "blurring function," which tends to merge individual reflection arrivals.

The wavelet can be considered a digital filter that converts the reflectivity to a seismic trace. To obtain reflectivity, we attempt to remove the filtering effects of the wavelet. This undoing of the convolution of the wavelet with the reflectivity is termed *deconvolution*. Let us now examine convolution and deconvolution in mathematical terms.

The seismic trace is a sum of delayed amplified wavelets that travel down into the earth and are reflected back to the surface with different time delays. On reflection, the wavelet is amplified by the reflection coefficient value. The digital convolutional model for the seismic trace is given by

$$x_t = \sum_\tau w_\tau r_{t-\tau} , \qquad (6)$$

where $x_t$, $w_t$, and $r_t$ represent time sequences for the seismic trace, wavelet, and reflectivity sequences, respectively. Symbolically, we can rewrite equation 6 as

$$x_t = w_t * r_t . \tag{7}$$

To recover the reflectivity perfectly, we would design a digital filter, $f_t$, so that

$$f_t * w_t = \delta_t ; \tag{8}$$

hence, we can filter, or "deconvolve," the seismic trace to obtain the reflectivity

$$f_t * x_t = r_t . \tag{9}$$

Much has been written about deconvolution, including books by Claerbout (1976), Robinson and Treitel (1980), and Yilmaz (2001). Ideally, deconvolution would provide the perfect answer to resolution problems. However, its performance often falls short of perfection, largely because of noisy signals, the band-limited nature of the seismic wavelet, and our lack of knowledge about the wavelet's form. Although a deconvolution filter will not perfectly shape the seismic wavelet to an ideal delta function because of the finite band-pass of our data, we can design filters that will shape wavelets to band-limited zero-phase versions of a spike.

Of course, the deconvolution process presupposes that we know the seismic wavelet, although this is seldom if ever the case. Nevertheless, deconvolution can significantly improve the high-frequency content of data. An example of this is shown in Figure 4. The Lower Cretaceous wedge in the input data is resolved more clearly after deconvolution. This is achieved by applying a Wiener deconvolution filter, which shapes the wavelet to a narrower, more symmetrical output. This spiking deconvolution commonly is applied to seismic data to improve the temporal resolution of closely spaced reflected arrivals.

## Lateral resolution

We have just visited the question of vertical resolution, that is, resolving thin beds by increasing the high-frequency content of data. It is natural also to consider resolution in a lateral sense. This problem could be posed by another question: How small is a small hole in a reflector? In other words, how small a spatial anomaly can we hope to resolve with our seismic data? This relates to the question of Fresnel zones and the excellent description given by Lindsey (1989).

**Figure 4.** Deconvolution example (from Lines and Treitel, 1984) illustrating (a) input data, (b) seismic wavelet, (c) deconvolved wavelet, and (d) deconvolved data.

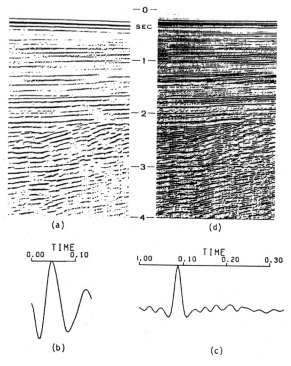

As in the case of vertical resolution, in which we examined the interference of closely spaced reflections, let us follow Lindsey's discussion and examine closely spaced reflections from a disk. Consider a reflecting plane with a hole of Fresnel diameter, $F_D$. Figure 5 shows a vertical cross section through the disk with raypaths from the center and edges of the disk. The arrivals from the center and edge of the disk become indistinguishable when their paths differ by a certain distance. If we use the limits of resolution based on the Widess criterion and advocated by Berkhout (1984), this distance difference is $\lambda/8$. If we use the criterion based on Rayleigh's analysis and advocated by Sheriff (1980), this difference is $\lambda/4$. It is not worth expending too much energy in the debate of which criterion to use; the principle behind the Fresnel zone will be the same.

Let us do the analysis. If we use Pythagoras' theorem for Figure 5 and Berkhout's criterion, we see that the relationship defining the Fresnel diameter is given by

$$(Z + \frac{\lambda}{8})^2 = Z^2 + (\frac{F_d}{2})^2 \tag{10}$$

or

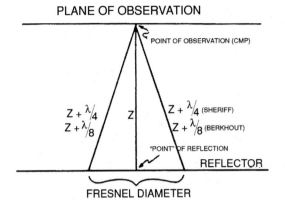

PLANE OF OBSERVATION

POINT OF OBSERVATION (CMP)

$Z + \lambda/4$   $Z$   $Z + \lambda/4$ (SHERIFF)
$Z + \lambda/8$      $Z + \lambda/8$ (BERKHOUT)

"POINT" OF REFLECTION

REFLECTOR

FRESNEL DIAMETER

**Figure 5.** Illustration (from Lindsey, 1989) explaining the geometry for determinations of Fresnel zones.

$$\frac{\lambda Z}{4} + \frac{\lambda^2}{64} = \frac{F_d^2}{4} . \tag{11}$$

If the wavelength is small relative to depth, i.e., $\dfrac{\lambda}{Z} \ll 1$, we can eliminate higher-order terms to give a good approximation for the Fresnel diameter:

$$F_d \cong \sqrt{\lambda Z} . \tag{12}$$

From equation 12, we see that the Fresnel diameter is given by the square root of the product of wavelength and depth. (If we use Sheriff's criterion rather than Berkhout's, then $F_d = \sqrt{2\lambda Z}$ .) In other words, spatial resolution degrades with increasing depth and seismic wavelength. For example, a Fresnel diameter for a reflector depth of 100 m and a wavelength of 64 m would be 80 m. A Fresnel diameter for a reflector depth of 900 m and a wavelength of 100 m would be 300 m.

Large Fresnel diameters, such as in the latter case, are quite disconcerting when we consider resolution. What is to be done to alleviate the problem? There are two possibilities — reduce wavelength or reduce depth. The reduction of wavelength can be tackled in the same manner as in the case of vertical resolution. We can attempt to reduce wavelength by increasing the high-frequency content of our data. Deconvolution should help, although it is usually less effective than in the case of enhancing vertical resolution.

The other alternative is to reduce the depth to our reflection target. This is not as absurd as it might seem. Seismic migration can be viewed as a seismic processing tool that continues our sources and receivers downward to the level of the seismic reflector. This is the migration concept developed by Claerbout (1976), among others. Therefore, an ideal migration would effectively set $Z = 0$ in equation 11. This would mean that the Fresnel diameter would be given by $F_d = \frac{\lambda}{4}$. For most exploration examples, this causes a substantial improvement in Fresnel diameters.

If we revisit the case of the 900-m reflection target and 100-m Fresnel diameter, an ideal migration would reduce the Fresnel diameter to 25 m, which would be one-twelfth, or less than 10%, of the Fresnel diameter for unmigrated data. The collapsing of Fresnel zones and the concomitant improvement of spatial resolution is an important bonus of seismic migration. Migration may not always give such a spectacular improvement in resolution, as in this simple example, because our velocity estimates are imperfect; consequently, our migrations are imperfect. Nevertheless, we can sometimes see some unexpectedly beneficial results of migration on "flatland" data because of collapsing Fresnel zones.

The resolution enhancement caused by migration becomes obvious when one examines data examples such as the Gregoire Lake 3D survey shown by Pullin et al. (1987). This case history is discussed later, in our chapter on heavy oil, but let us do the Fresnel-zone calculations for this case. The problem involved seismic imaging of a Devonian reflector at 250 m with seismic wavelengths that were about 10 m. Although this wavelength may be a bit optimistically compact, the velocity was about 2000 m/s and the data did contain frequencies in the range of 25–240 Hz.)

Prior to migration, the diameter of the Fresnel zone would be 50 m. After migration, the Fresnel zone ideally could be collapsed to 2.5 m, with a resolution improvement of 20. If one examines the detail on the 3D migrations of the Devonian channel reflection in Pullin et al. (1987), these improvements in resolution predicted by Fresnel-zone calculations are not out of the realm of possibility. In fact, migration itself can be viewed as a "spatial deconvolution."

In the interpretation of thin beds, i.e., those that are a fraction of a seismic wavelength, seismic resolution is crucial. Resolution can be enhanced by decreasing the seismic wavelength and collapsing Fresnel zones. This can be achieved by a combination of acquisition and processing techniques. In acquisition, high-frequency content generally can be increased by the use of small charges and buried geophones. In processing, we can improve resolution by the proper use of deconvolution and migration.

The material below is a revised version of a *Canadian Society of Exploration Geophysicists Recorder* paper by Lines et al. (2001) dealing with depth migration. We hope it will enhance the spatial resolution of seismic arrivals.

# Depth imaging — "If we could turn back time"

Seismic migration is an essential processing step in exploration plays involving any structural complexity, because migration is the process of placing seismic-reflection energy in its proper subsurface location. The popular song "If I Could Turn Back Time" can describe the process of achieving accurate migration of seismic data through backward time propagation of wavefields.

In the process of imaging complex structures such as faults and folds, we seek to reverse the path of surface-recorded seismic reflections back to the geologic reflectors. In essence, the idea is to "depropagate" seismic waves. The earliest geophysical publication on the use of complete wave equation solutions for reverse-time migration was presented by Hemon (1978) in a French paper titled "Equations d'onde et modeles." Hemon published the paper but did not believe it was practical.

At about the same time, Dan Whitmore of Amoco Research was using the method extensively for the very practical problems of imaging overthrust structures and salt domes, but he did not publicly disclose his results until he participated in the migration workshop at the 1982 SEG annual meeting. Shortly thereafter, several papers on the subject appeared, including independent research by Baysal et al (1983), Loewenthal and Mufti (1983), McMechan (1983), and Whitmore (1983).

Baysal's paper initiated the term *reverse-time migration* and gave some interesting overthrust examples. McMechan's paper gave a lucid description of the algorithm's simplicity and its generality. Subsequently, McMechan and his students have adapted reverse-time migration to almost every combination of migration problems, including 2D, 3D, poststack, prestack, isotropic, and anisotropic situations. Whitmore et al. (1988) and Zhu and Lines (1998) showed how reverse-time migration compared with other popular depth-imaging methods, such as Kirchhoff and phase-shift migration. Mufti et al. (1996) showed that with variable grids for finite-difference calculations, the method could be practically applied in three dimensions to the depth imaging of salt domes in Gulf Coast exploration.

# Space and time

Let us investigate the principles of reversing time in the migration process. Although some modern physicists will disagree, for all practical purposes, we live in a 4D world with three spatial dimensions plus the dimension of time. Any event that occurs is specified by its location in 3D space and a specific time of occurrence. Physics deals with the 4D space-time continuum. In space, we can go in forward or reverse directions. For example, in a 3D Cartesian system, we can travel north or south, east or west, up or down. However, with time in the physical world, we can progress in only one direction. We cannot physically go back in time and violate the principle of causality.

How is it possible, then, to perform reverse-time migration? To achieve this, we need to revisit the wave equation, which defines the propagation of seismic energy. Recall solutions to the wave equation for a 1D homogeneous medium given by

$$\frac{\partial^2 u}{\partial x^2} = \frac{1}{v^2}\frac{\partial^2 u}{\partial t^2}. \tag{13}$$

D'Alembert's principle states that a general solution is of the form $u = f(x+vt) + g(x-vt)$, where $f$ and $g$ are arbitrary functions that are twice differentiable. This can be shown by using the chain rule in partial differentiation and is derived in many books on the physics of waves, including, for example, Becker (1954). It is fascinating that the solutions to the wave equation remain valid if we change the sign of $t$, the time variable. In other words, the underlying physical processes of waves and the mathematical solutions to the wave equation remain unchanged if time is reversed.

# Time-reversed acoustics

This property of waves has led to the field of time-reversed acoustics, as described in a *Scientific American* article (November 1999) by Mathias Fink, director of the Waves and Acoustics Laboratory in Paris. At this laboratory, an array of microphones and speakers produces an acoustical "time-reversal" mirror. A speaker pronouncing a message "Bonjour" to the mirror would receive the message "Ruojnob" almost instantaneously converging back to the speaker's mouth, as if the sound wave had experienced time reversal. This is achieved by microphones detecting the sound wave, passing the signal to a computer that stores the wave, reverses the signal, and sends it back along its original path. Although this may seem to be a somewhat comical experiment, it has many practical implications in the fields of seismic migration, underwater communication, materials testing, and medicine.

In medicine, for example, time-reversed acoustics can be used to break up kidney stones. Ultrasonic pulses from transducers bouncing off kidney stones can be focused back on the kidney stones, causing them to break up. There are also research efforts into using the time-reversed acoustics method for the destruction of tumors. Many useful and interesting applications are described in Fink's excellent *Scientific American* article, although the seismic applications are not mentioned. This missing item is unfortunate, because seismic migration was probably one of the first successful applications of the method.

## Reflection seismology, exploding reflectors, and reverse-time migration

In the simplest of terms, we can relate reflection seismology to reverse-time migration in terms of seismic wavefield propagation. If we can imagine a movie showing wave propagation of reflected seismic energy, migration could be achieved by simply running the movie backward so that seismic energy travels back to the reflectors from which it came. In a sense, migration is indeed the inverse of seismic wavefield modeling. A simple explanation for the zero-offset seismic experiment is given by Figure 6, from an *AAPG Explorer* article by Bording and Lines (2000).

The reflection seismic experiment is shown in Figure 6a for a source and receiver at basically the same surface location. Seismic waves propagate with velocity *v* from a source (at the flag) down to a reflector and then back to a receiver (at the triangle). Loewenthal et al. (1976) pointed out that we can view this experiment differently, in terms of an "exploding-reflector" model, which is described in Figure 6b. For most geometries, we also can think of the seismic response as having been generated by seismic waves that travel a one-way path from "exploding reflectors" to the surface receivers (or half the distance) at a velocity that is half the true seismic velocity.

The "exploding-reflector model" is a very useful and economical concept. It closely approximates zero-offset stacked section with considerably less compute time than computing individual shot gathers. It also helps us

**Figure 6.** Zero-offset seismic experiment (modified from Bording and Lines, 2000). Diagrams represent the chain of operations involved in a zero-offset seismic experiment, from the seismic reflection experiment in the upper left to the exploding-reflector model to the application of reverse-time migration.

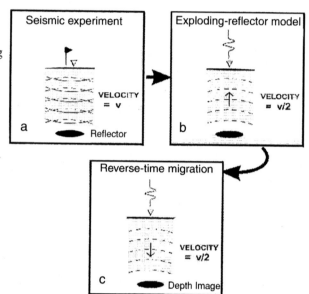

to understand depth imaging, because reverse-time migration processes the seismic data by reversing the path of the exploding-reflector model, as shown by Figure 6c. The image is obtained by propagating the reflected energy backward in time with velocity $v/2$ to its point of origin (the reflector). For this discussion, we have dealt with the zero-offset case of coincident sources and receivers — an approximation to the case of poststack migrations. All the migration methods (reverse-time, Kirchhoff, phase-shift, finite-difference, etc.) have prestack versions that lend themselves to greater generality. We shall describe very briefly the imaging principle for prestack migration, referring the reader to Yilmaz (2001) for a discussion of the details of this important technology.

The exploding-reflector model of Loewenthal et al. (1976) allows us to think of a stacked section, which approximates a zero-offset section, in terms of the response to a set of reflectors that exploded at time zero. This thought experiment allows us to formulate the migration problem as that of imaging the recorded wavefield at time zero, i.e., when and where the reflectors exploded. In fact, the exploding-reflector model provides a very natural motivation for reverse-time migration. ("I have a section with positive times. How do I get the corresponding section inside the earth at time zero? I have to propagate a wavefield with the arrow of time reversed!")

Unfortunately, this model doesn't tell us how to track migration. Reflection inside the earth takes place *after* the source has been excited but *before* the reflected energy has been recorded. Therefore, the time of reflection lies between time zero (the source time for a given shot) and the recording time for that event. If, at a particular location inside the earth, we find the time when the wavefield from the source has passed through that location, we know that reflection has occurred there at that time. Then we must evaluate the recorded wavefield at the reflector location at the time of reflection and compute the time from the receiver to the reflector. The first of these times expresses, in a sense, the downward continuation of the source wavefield, and the second expresses the downward continuation of the receiver wavefield. If these times are stripped away, then both wavefields have been continued downward all the way to the reflector, and reflection occurs at time zero.

## Example from the Canadian Foothills

The usefulness of "time-reversed acoustics" in depth imaging for foothills exploration can be demonstrated by examining some fault-fold models typical of the Canadian Foothills. The example shown in Figure 7 requires the interpreter to distinguish between a fault-bend fold model and a fault-propagation fold model. The total length of each model is 10 km with a maximum depth of 4200 m. Cell sizes are 10 m × 10 m. Velocities in Figure 7 are 2000 m/s (light gray), 2500 m/s (medium gray), 4000 m/s (dark

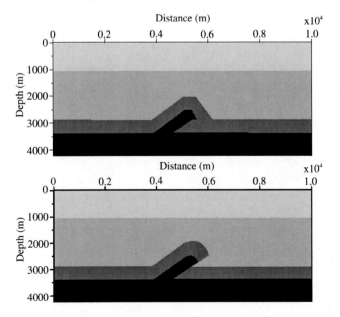

**Figure 7.** Models used in the study illustrated in Figure 6. Both models represent common structures seen in foothills environments. The upper model represents a fault-bend fold; the lower model is a fault-propagation fold.

gray), and 5500 m/s (black). No topography has been included, layer thicknesses have been preserved, and the overall amount of structural shortening in each model is equivalent. Therefore, the only variable in each model and resulting image is the structural geometry.

Although these models appear to be quite similar, they represent different structural styles that might influence trapping mechanisms and reservoirs in oil and gas targets. The difficulty in imaging steep dips is the definition of the forelimbs for these models. This problem has plagued geophysicists and interpreters for many years. As shown in Figure 8, the two unmigrated seismic responses show little resemblance to the actual models except where the layers are flat. The power of migration is its ability to move this surface-recorded energy back into its proper depth location. We will assume for the moment that we have perfect estimates of the velocity field. This is a big assumption, because velocity estimation itself is a major problem (perhaps *the* major problem for prestack migration). Zhu et al. (1998) have shown how we can use migration itself to estimate migration velocities.

The depth images in Figure 9, examples of fault-bend folds and fault-propagation folds, show that migrations can unravel the unmigrated seismic sections and produce depth images that do clearly distinguish between the two models.

It is impressive how the migration process can move reflection energy to its correct position in the subsurface. However, if one views this in terms of time-reverse acoustics, perhaps the process is not mysterious. After all,

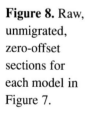

**Figure 8.** Raw, unmigrated, zero-offset sections for each model in Figure 7.

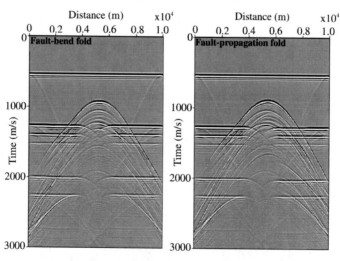

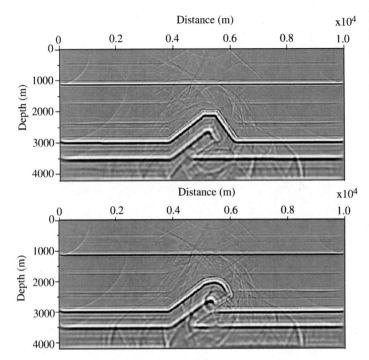

**Figure 9.** Reverse-time migrated sections of each model in Figure 7.

if waves can be moved backward in time, we can move the seismic arrivals back to their reflection points.

Time-reversed acoustics has found applications in many fields in addition to seismology. One wonders how many other applications await us if we could turn back time in wave propagation.

# References

Baysal, E., D. D. Kosloff, and J. W. C. Sherwood, 1983, Reverse-time migration: Geophysics, **48**, 1514–1524.

Becker, R. A., 1954, Introduction to theoretical mechanics: McGraw-Hill Book Co.

Berkhout, A. J., 1984, Seismic resolution: Geophysical Press, Pergamon.

Bording, R. P., and L. R. Lines, 2000, Seismic modeling makes waves: AAPG Explorer, **21**, 38–41.

Claerbout, J. F., 1976, Fundamentals of geophysical data processing: McGraw-Hill Book Co.

Fink, M., 1999, Time-reversed acoustics: Scientific American, November, 91–97.

Hemon, C., 1978, Equations d'onde et modeles: Geophysical Prospecting, **26**, 790–821.

Jenkins, F. A., and H. E. White, 1957, Fundamentals of optics: McGraw-Hill Book Co.

Kallweit, R. S., and L. C. Wood, 1982, The limits of resolution of zero-phase wavelets: Geophysics, **47**, 1035–1046.

Lindsey, J. P., 1989, The Fresnel zone and its interpretive significance: The Leading Edge, **10**, 33–39.

Lines, L. R., and S. Treitel, 1984, Tutorial: A review of least-squares inversion and its application to geophysical problems: Geophysical Prospecting, **32**, 159–186.

Lines, L. R., S. Rushton, D. C. Lawton, and S. H. Gray, 2001, Depth imaging — "If we could turn back time": Canadian Society of Exploration Geophysicists Recorder, **9**, 28–31.

Loewenthal, D., and I. R. Mufti, 1983, Reversed time migration in the spatial frequency domain: Geophysics, **48**, 627–635.

Loewenthal, D., L. Lu, R. Roberson, and J. Sherwood, 1976, The wave equation applied to migration: Geophysical Prospecting, **24**, 380–399.

McMechan, G. A., 1983, Migration by extrapolation of time-dependent boundary values: Geophysical Prospecting, **31**, 413–420.

Mufti, I. R., J. A. Pita, and R. W. Huntley, 1996, Finite-difference depth migration of exploration-scale 3-D seismic data: Geophysics, **61**, 776–794.

Pullin, N., L. Matthews, and K. Hirshe, 1987, Techniques applied to obtain very high resolution 3-D seismic images at an Athabasca tar sands thermal pilot: The Leading Edge, **12**, 10–15.

Ricker, N., 1953, Wavelet contraction, wavelet expansion: Geophysics, **18**, 769–792.

Robinson, E. A., and S. Treitel, 1980, Geophysical data processing: Prentice-Hall.

Sheriff, R., 1980, Nomogram for Fresnel zone calculation: Geophysics, **45**, 968–972.

Whitmore, N. D., 1983, Iterative depth migration by backward time propagation: 53rd Annual International Meeting, SEG, Expanded Abstracts, 827–830.

Whitmore, N. D., S. H. Gray, and A. Gersztenkorn, 1988, Two-dimensional depth migration: A survey of methods: First Break, **6**, 189–200.

Widess, M., 1973, How thin is a thin bed?: Geophysics, **38**, 1176–1254.

Yilmaz, O., 1987, Seismic data processing: SEG.

Yilmaz, O., 2001, Seismic data analysis: Processing, inversion, and interpretation of seismic data, 2 v.: SEG.

Zhu, J., and L. R. Lines, 1998, Comparison of Kirchhoff and reverse-time migration methods with applications to prestack depth imaging of complex structures: Geophysics, **63**, 1166–1176.

Zhu, J., L. R. Lines, and S. H. Gray, 1998, Smiles and frowns in migration/velocity analysis: Geophysics, **63**, 1200–1209.

# Chapter 7

# Aliasing for the Layperson

This chapter consists of a slight modification of a paper by Lines et al. (2001) published in the *Canadian Society of Exploration Geophysicists Reporter*. It is included in this text with permission.

## Introduction — Aliasing in everyday life

Examples of aliasing can be observed in western movies by watching the motion of stagecoach wheels. As the stagecoach starts to move, we observe its wheels rotating in the expected forward direction. As the stagecoach speeds up, we see that the wheels appear to rotate in the opposite direction to the initial one. We realize that this phenomenon happens because the movie consists of a number of frames, or "digital samples." With the wheels' acceleration, the digitized succession of frames will show the spokes appearing to move opposite to the actual direction of rotation. This occurs because the movie camera has undersampled the wheels' motion. Similar effects in real life occur when one watches helicopter blades as they speed up. Our brains sample the apparent motion too slowly to detect the actual motion, and the blades may appear to have a slower rate of rotation than the true speed.

## Strobe-light experiments, sampling, and the Nyquist frequency

An illuminating example of aliasing and the sampling concepts comes from a physics experiment in which we attempt to detect the angular speed of a rotating disk by use of a flashing strobe light. Note that if the frequency of the strobe flashes coincides with the frequency of the disk's rotation, we can "freeze" the motion of a point on a disk. However, the freezing of a point on the disk also occurs if the disk's rotational frequency is at some integral multiple of the strobe light's frequency.

For example, consider the disk in Figure 1, which is rotating clockwise at 1 cycle per second. We mark a point on the disk with an arrow to indicate its position. If we set our strobe flashing at periods of 1, 2, and 3 seconds (i.e., at frequencies of 1, $1/2$, and $1/3$ Hz), we will see the arrow appear at the same location. The highest strobe frequency, which allows the arrow to be frozen, is the sampling frequency of 1 Hz. If we set the strobe light to flash slightly faster than once per second, we see that the arrow appears to move counterclockwise. For example, if the strobe flashes every 0.75 s, we first see the arrow appear at −90° (or +270°), then at 180°, and then at 90° from its initial vertical position. If we set the strobe light to flash every 0.5 s (2 Hz), we see the arrow alternate between pointing vertically upward and downward. The strobe light samples the disk's position twice every cycle.

Setting the strobe light at a sampling rate of 0.5 s, we can detect the 1-Hz rotation of the disk. For a strobe-light sample interval of 0.5 s, the maximum detectable frequency of disk rotation is 1 Hz. If the disk rotates any faster than 1 Hz, we must use a finer sampling interval to detect it. This maximum measurable frequency for a given sample interval is known as the

**Figure 1.** Flashing strobe light on a rotating disk.

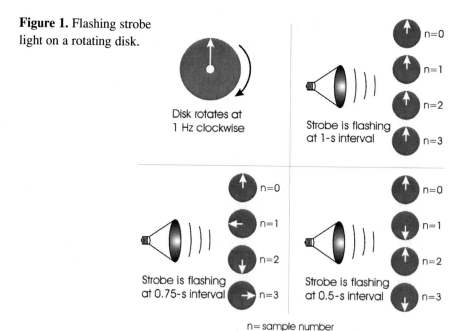

Disk rotates at 1 Hz clockwise

Strobe is flashing at 1-s interval

$n=0$
$n=1$
$n=2$
$n=3$

$n=0$
$n=1$
$n=2$
$n=3$
Strobe is flashing at 0.75-s interval

$n=0$
$n=1$
$n=2$
$n=3$
Strobe is flashing at 0.5-s interval

$n=$ sample number

Nyquist frequency, in honor of Dr. Harry Nyquist, an American physicist and engineer.

Through Fourier analysis, any seismic signal can be decomposed into weighted sums of sines and cosines. By analogy with the rotating disk and strobe light, it is useful to note that we must sample the sine wave at least twice every cycle (sampling a peak and a trough) to detect its frequency. Figure 2 shows the consequence of inadequate time sampling; we see a sinusoidal wave at a lower frequency. For example, if we measured the sine wave only once per cycle, we would measure the same sample value, and it would appear as a DC (constant-amplitude) signal. In other words, if our sample interval is set at $\Delta t$, then the Nyquist frequency is given by

$$F_N = \frac{1}{2\Delta t}. \tag{1}$$

In seismic recording, mechanical vibrations as recorded by seismometers or hydrophones ultimately are converted into a sequence of digital values. If we set a sample interval at a typical value of 2 ms, the Nyquist frequency is 250 Hz. Frequencies above 250 Hz in our signals will be aliased or recorded as apparently lower frequencies, just as in the case of the wagon wheels or rotating disks that appear to be rotating in the wrong direction. Therefore, to avoid such problems, the antialias filter in the recording truck must be set to electronically filter out frequencies above 250 Hz in analog signals prior to digitization. This is a necessary step in seismic recording to avoid the temporal aliasing encountered by digital sampling of time series.

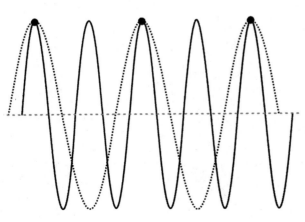

**Figure 2.** Because of inadequate sampling, the sinusoidal wave is displayed at a lower frequency than is actually the case. Adapted from Cary (2001).

● Sampling point

# Spatial aliasing

Although there is an analog antialias filter in time sampling, the same type of filter does not exist for spatial sampling, in which we record with a discrete number of receivers. To appreciate the problem of spatial aliasing, we refer to an example revised from Hatton et al. (1986). Figure 3 shows a set of 40-Hz sinusoidal traces dipping to the right with delays of 4 ms per trace. In the second panel, with delays of 10 ms/trace, the traces still appear to dip to the right. However, if the traces are delayed by 20 ms per trace, they appear to be dipping to the left. Aliasing has occurred.

Why does this happen? Recall that temporal sampling requires that we sample a sinusoid twice every period. In other words, adjacent samples must be less than half a period apart. If we observe the spatial samples of adjacent traces at a given time, we note that aliasing occurs when the delay between traces (20 ms in the aliased example) is greater than half of the period, which in this case is 12.5 ms. The increased delay per trace causes the aliasing. The delay, or "stepout" difference, between the aliased and unaliased sections could occur for at least two reasons. If the events represent reflections, the stepout difference may be a result of the physical dip of the reflector. Alternatively, it could be a result of the spacing between geophones. In other words, this differencing could occur if the geophone spacing in the aliased section was twice that in the unaliased section. In that case, the geophone sampling may be too sparse to sample the wavefield adequately.

The example in Figure 3 suggests that insufficient spatial sampling and steep dip will cause spatial aliasing. Spatial aliasing also is affected by the

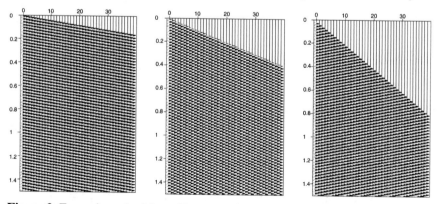

**Figure 3.** Example revised from Hatton et al. (1986) showing how increased arrival delays create aliasing.

temporal frequency of signals. Figure 4 is another example that demonstrates this phenomenon. The left part of Figure 4 shows the unaliased case of 40-Hz sinusoids with a stepout of 10 ms/trace. If we change these sinusoids from 40 Hz to 60 Hz and maintain the same stepout per trace, we also can see an apparent dip to the left instead of an apparent dip only to the right. Aliasing is caused by increased frequency.

To further analyze spatial aliasing, let us consider the situation in Figure 5 of an emerging plane wave traveling upward with a velocity $v$ and hitting the surface at an angle $\theta$. Let the spacing between geophone elements be $\Delta x$. To avoid aliasing, we recall that the time stepout between adjacent traces

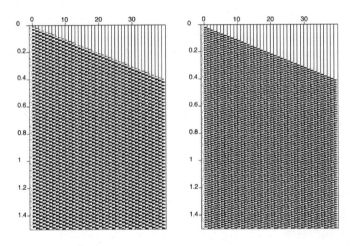

**Figure 4.** Example revised from Hatton et al. (1986), showing how increased frequency creates aliasing.

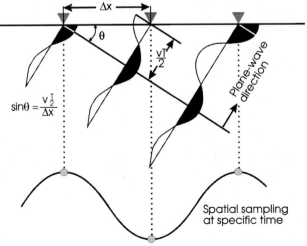

▼ = Geophone

$$\sin\theta = \frac{v\frac{T}{2}}{\Delta x}$$

**Figure 5.** Derivation of aliasing criteria.

should be, at most, half the period of the sinusoid, which we denote by $T/2$. If we examine Figure 5, we see by inspection that $\sin\theta = vT/2\Delta x$.

We can relate this analysis of a transmitted plane wave to the case of a reflected wave if we recall the "exploding-reflector" model for zero-offset data described by Loewenthal et al. (1976). Those authors pointed out that reflections resulting from two-way travel from a point to a reflector and back to the same point can generally be described by a wave experiencing one-way travel from the reflector to the source point if the wave travels at half the velocity of the medium. If we consider the emerging plane wave in Figure 5 to be from an exploding reflector, we can replace $v$ with $V/2$ in the previous expression for $\sin\theta$, where $V$ is the seismic velocity of the medium. We now have the expression that gives the limiting values to avoid aliasing. If the frequency of our sinusoidal waves is fixed at $f = 1/T$, then the geophone spacing, $\Delta x$, must satisfy the following condition, which we obtain by substituting for the medium velocity and period in the previous expression:

$$\Delta x \le \frac{V}{4f\sin\theta}. \tag{2}$$

(This expression is also given by Yilmaz, 2001.)

Equivalently, if our geophone spacing is held constant, we must limit the temporal frequencies or the dip by the above expression to avoid aliasing. We should note that equation 2 is somewhat simplified, in that it is derived for a zero-offset section and is generally not the relationship for geophone spacing in typical shot gathers. For more general cases, refer to Cordsen et al. (2000, p. 21–23) or to Cary and Li (2001).

In practical terms, we need to be aware of how this expression affects our survey design and the temporal frequencies that we can hope to image in our seismic experiments. Aliasing can be avoided in time by electronically filtering out the high frequencies above the Nyquist frequency prior to digitization. How can we avoid the effects of spatial aliasing?

If we examine equation 2, we see that one way to avoid aliasing is to design the seismic survey with closely spaced geophone groups so that equation 2 is satisfied. This requires that we have some knowledge of the wave's velocity, the steepest dip of the reflecting horizons, and the largest usable temporal frequency. Of course, there is a practical trade-off between the expense of a survey (which increases with decreasing group spacing) and our desire to avoid spatial aliasing.

Sometimes we do not have complete knowledge of steepest dips and minimum velocity a priori, and we are dealing with data that have been recorded with geophone spacings that allow aliasing of high-frequency events. What do we do in such situations?

Some remedies to this problem are outlined by Yilmaz (2001). One approach is to decrease the temporal frequencies by low-pass filtering of our data. This approach may suffice if we are not concerned about the resolution of thin beds. However, removing the high frequencies will increase the effective wavelength and will degrade resolution. Another approach is to decrease the apparent dip of our events by time shifting, but because our beds generally have differing dips, this may decrease the dip of some beds while increasing that of others. One of the best approaches is the use of intelligent interpolation routines. If we can interpolate data successfully, in some cases we could produce data effectively with half the original group spacing. For the aliasing example in Figure 3, we can interpolate the aliased traces that have stepouts of 40 ms/trace to produce traces that have stepouts of 20 ms/trace.

Seismic-trace interpolation can be a worthwhile method for improving the migration of spatially aliased data. Any successful interpolation of aliased data must provide additional information about the wavefield if it is to overcome the missing data problem. One such interpolation method was demonstrated by Spitz (1991), who effectively applied a multichannel f-x domain interpolation by constraining the wavefield at each frequency to consist of a small number of linear dipping events, which may or may not be aliased. The algorithm automatically determines the dip and amplitude of these events within small overlapping data windows. Porsani (1999) produced a modified version of Spitz's algorithm using prediction filters. This intelligent interpolation of seismic traces generally is accomplished by applying multichannel filtering with appropriate constraints.

# Example of "back-of-the-envelope" calculations

One of the best ways to understand how we can deal with aliasing is with a simple example. Consider a seismic survey in which the following characteristics are expected:

| | |
|---|---|
| Medium velocity | $V = 2500–5000$ m/s |
| Frequency bandwidth | $f = 10–100$ Hz |
| Structural dips | $\theta = 0–40°$ |

First, let us examine the required time sampling, $\Delta t$, to avoid temporal aliasing. We want to ensure that the maximum frequency, 100 Hz ($F_N$), is sampled properly. Applying equation 1, we find

$$\Delta t \le \frac{1}{2F_N} = \frac{1}{2\,(100)} = 5\,\text{ms}\,. \tag{3}$$

Therefore, a time-sample interval of 5 ms or smaller is required. Most conventional surveys are recorded at 4 ms or less, so there is little concern for temporal aliasing.

Let us examine the required geophone spacing, $\Delta x$, to avoid spatial aliasing. We must consider the limiting case in which $V = 2500$ m/s, $f = 100$ Hz, and $\theta = 40°$. Applying equation 2, we find

$$\Delta x \le \frac{V}{4f \sin \theta} = \frac{2500}{4(100) \sin 40°} = 9.7\,\text{m}\,. \tag{4}$$

Therefore, a maximum geophone spacing of 9.7 m (essentially 10 m) is required to image all expected dips for the given medium velocity and bandwidth. If the spacing is not feasible, processing techniques such as filtering or trace interpolation, as discussed above, may be necessary.

# Summary and future possibilities

An effective interpolation between traces can allow us to avoid spatial aliasing problems, but of course, it generally would not be as good as reliable recorded signals. Nevertheless, if we can reduce acquisition costs and, at the same time, maintain the integrity of processed seismic data by avoiding aliasing, there is considerable practical use for interpolation. In the future, we may see that neural networks will prove to be a useful method of trace interpolation.

This chapter is tutorial and has related aliasing to everyday examples, while exploring some fundamentals of temporal and spatial aliasing.

# References

Cary, P., 2001, Temporal and spatial aliasing: Canadian Society of Exploration Geophysicists Recorder, **26**, no. 2, 20.

Cary, P., and X. Li, 2001, Some basic imaging problems with regularly-sampled seismic data: Presented at the 71st Annual International Meeting, SEG.

Cordsen, A., M. Galbraith, and J. Peirce, 2000, Planning land 3-D seismic surveys: SEG.

Hatton, L., M. H. Worthington, and J. Makin, 1986, Seismic data processing — Theory and practice: Blackwell Scientific Publications.

Lines, L., K. Brittle, I. Watson, and P. Cary, 2001, Aliasing for the layperson: Canadian Society of Exploration Geophysicists Recorder, **26**, no. 4, 10–16.

Loewenthal, D., L. Lu, R. Roberson, and J. W. C. Sherwood, 1976, The wave equation applied to migration: Geophysical Prospecting, **24**, 380–399.

Porsani, M. J., 1999, Seismic trace interpolation using half-step prediction filters: Geophysics, **64**, 1461–1467.

Spitz, S., 1991, Seismic trace interpolation in the f-x domain: Geophysics, **56**, 785–794.

Yilmaz, O., 2001, Seismic data analysis: Processing, inversion, and interpretation of seismic data: SEG.

# Chapter 8

# Seismic Ties from Well Data

Reflection seismology is essential for modern petroleum exploration. It allows us to estimate the changes in geologic formations between control points of existing wells. In this sense, seismic data allow us to interpolate between wells.

Conversely, this interpolation is possible only if we can identify seismic reflections. This is done by using correlations to well data. Although formation depths, resistivity, SP, and gamma-ray logs are useful, sonic and density logs provide the most useful information for correlating with seismic data. This correlation often is done with synthetic seismograms. Synthetic seismograms of primary reflections usually can be considered as band-limited reflection coefficients. These coefficients are dependent on contrasts in acoustic impedance (product of density, $\rho$, and P-wave velocity, $v$), that is, the reflection between two layers (say layer 2 and layer 1) is given by

$$R = \frac{\rho_2 v_2 - \rho_1 v_1}{\rho_2 v_2 + \rho_1 v_1} . \tag{1}$$

Therefore, the density and velocity logs of wells are used to calculate a set of reflection coefficients, which is then filtered so that the dominant frequencies match those in the surface seismic data. The set of reflection coefficients is then band limited to the same frequency band as the seismic data. The formations are identified on the synthetic seismogram by using well-log formation information. Correlations are then made between the reflections on the well-log synthetic seismograms and those on the seismic data. Several examples appear in this book, including examples in Chapter 15.

Figure 1 shows an example of such correlations for traces from a Grand Banks seismic line. In the reflection arrivals, one can see similarities between the synthetic trace and delayed arrivals in the field data. For example, the peak at 0.42 s on the synthetic trace matches the peaks at 0.46 s on the field traces.

## Why are well ties imperfect?

Generally, surface recordings of seismic data are in the 5–100-Hz range, and sonic logs are in the 10–12-kHz range for monopoles sources. If the velocity were not dispersive (not frequency dependent), we might expect nearly perfect correlations. In some cases, as in the examples in Chapter 15, the correlations are very good and worthwhile.

However, in other cases, this match is not perfect. There are several reasons for this, including:

1) velocity dispersion

2) problems with sonic logs, such as cycle skipping or mud invasion

3) noisy seismograms

4) nonzero-phase wavelets

5) polarity differences between synthetic seismograms and seismic data

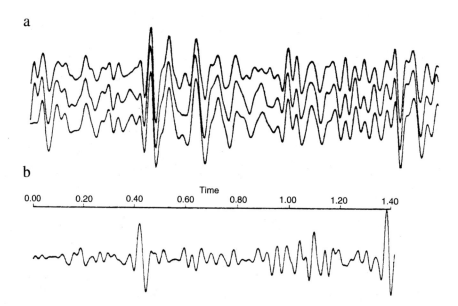

**Figure 1.** Comparison of (a) seismic traces with (b) a synthetic seismogram from a nearby well (from Lines and Treitel, 1985).

# Wavelets, well logs, and Wiener filters

The correlation between synthetic seismograms (band-passed reflection coefficients) and seismic arrivals usually assumes that wavelets are symmetrical and that the wavelet peak has zero delay in time. In signal processing, such wavelets are termed "zero phase."

Figure 1 shows an example of delays between wavelet arrivals compared to synthetic seismogram reflections. Although the seismic wavelet is usually unknown, we can estimate the wavelet from the seismic data by the use of well logs. A description of this estimation process is given by Lines and Treitel (1985). The process uses both seismic and well-log data.

In Chapter 5, we show that in the frequency domain, the Fourier transform of the seismic trace, $Y(f)$, is given by the product of the wavelet transform and the reflectivity transform:

$$Y(f) = W(f) \, R(f).$$

We record the trace to give $Y(f)$ and compute $R(f)$ from synthetic seismogram computation. The wavelet $W(f)$ is given by $W(f) = Y(f)/R(f)$. We then can design a Wiener (least-squares) filter, which deconvolves this wavelet throughout the seismic line. This process ideally should eliminate problems 4 and 5 listed in the previous section. If we have reliable well logs and good seismic data, the process is an effective wavelet deconvolution.

Matching synthetic seismograms with seismic data is an essential part of the interpretation process. Several problems can be involved, but the correlation is essential for tracing seismic reflections throughout the subsurface.

# Reference

Lines, L. R., and S. Treitel, 1985, Wavelets, well logs and Wiener filters: First Break, **3**, 9–14.

# Chapter 9

# Character, Continuity, Correlatability, and Coherence

Much of seismic interpretation can be viewed as the process of identifying seismic reflecting horizons between well ties. In some sense, the seismic-interpretation process can be viewed as interpolation between control points at well locations. In a previous chapter, we established a methodology for using synthetic seismograms at well locations to identify seismic reflectors. In this chapter, we discuss the next step — using these control points on seismic sections to identify reflecting horizons between wells. To that end, we will discuss seismic wavelet character and continuity, and we will use coherency measures, such as crosscorrelation — hence, the title.

## Character

Seismic character is defined by Sheriff (1991, p. 37) as "the recognizable aspect of a seismic event; the waveform which distinguishes it from other events." Character is defined by the amplitude and phase of the waveform. The amplitude is given by the height of a waveform's peak or trough and is expected to be related to the reflection coefficient for a boundary. For example, we might examine the amplitude of a seismic reflection from the top of a seismic reef. Reflections from an interface between shale and tight limestone would be strong because of a large impedance contrast. As we move to the porous, dolomitized reef, impedance contrasts are less, and the reflection strength will decrease, or create a "dim spot." For carbonate reefs, such character changes are indicators of reef porosity.

In addition to the amplitude or magnitude of reflection events, it is important to know the energy distribution in a seismic arrival. This distribution is defined by the phase, or timing, of the seismic arrival. Phase changes in seismic arrivals can be the result of seismic source characteristics, geophone or instrument phase delays, or geologic layering. Therefore, it is important to know the cause of phase delays.

For example, suppose we believe that the seismic wavelet in our processed section should be zero phase or symmetric in its appearance. Now, suppose that the arrival on the seismogram has a 90° phase shift. As shown by the Widess analysis of thin beds in Chapter 6, this is because of a thin layer of high impedance sandwiched between layers of low impedance. This geologic situation causes a positive reflection coefficient to be followed closely by a negative reflection coefficient. This creates a "differentiator" of the seismic pulse, which creates a 90° phase shift.

On the other hand, we might have deceived ourselves into believing that our wavelet was zero phase and thereby have failed to account for phase shifts, instruments, or source ghosts for buried sources. It is crucial to know the wavelet phase in our seismic sections.

A specific example of phase shifts is the issue of wavelet polarity. (In this case, we want to differentiate between phases of 0° and 180°.) In other words, we want to know whether our reflection is a peak or a trough. There are many ways to flip polarity on a seismogram. It can be flipped by geophone hookup, instrument settings, processing, or plot parameters.

In addition to these complications are the issues of polarity conventions. Although the conventions are arbitrary, it is important to know whether a peak for a seismic wavelet on the seismogram represents an increase or decrease in seismic impedance. In fact, it would be worthwhile for every seismic-section label to show a graph of seismic-impedance increase with either a peak or a trough, so the interpreter knows which convention was used.

With the variety of ways to reverse polarity on a seismic section, it is sometimes difficult to be sure about polarity. One method that can help solve this problem is the use of well logs and synthetic seismograms.

If sonic and density logs from a well location are a true representation of the earth's structure, we can establish the amplitude and phase (and hence the polarity) of the seismic wavelet. This process is described in a paper titled "Wavelets, well logs, and Wiener filters" by Lines and Treitel (1985). The process deconvolves the reflectivity function from the seismic trace to provide a wavelet estimate. We briefly describe the procedure as follows:

Recall the definition of the seismic trace as the convolution of the wavelet with the reflectivity. In the frequency domain, this means that the

Fourier transform of the trace, $Y(f)$, can be expressed as the product of the wavelet's Fourier transform, $W(f)$, with the reflectivity's Fourier transform, $R(f)$:

$$Y(f) = W(f)\,R(f).\tag{1}$$

Therefore, because well logs allow us to compute the reflectivity, we can compute $R(f)$ at a well location. Because we also can compute the Fourier transform of the trace, $Y(f)$, we can compute an estimate of $W(f)$:

$$W(f) = \frac{Y(f)}{R(f)}.\tag{2}$$

This wavelet estimation by use of seismic data and well logs will allow us to iron out the ambiguities of wavelets and reflectivities that exist with a given seismic trace. Although we express the wavelet estimate in frequency-domain terms, we may want to do the wavelet estimation by using Wiener filters in the time domain (Lines and Treitel, 1985).

## Continuity

Seismic continuity results from visual recognition of similar waveforms that are aligned. This continuity generally relates to lateral consistency of a reflection across the seismogram. Situations in which reflections are continuous or parallel on a seismic section often are described as "railroad tracks" and indicate a continuity of acoustic impedance or lithology. Figure 1 dis-

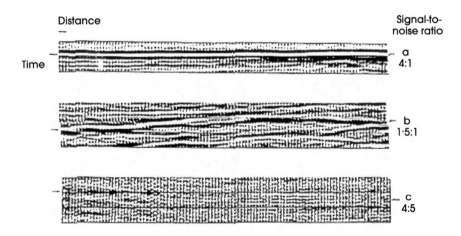

**Figure 1.** Different levels of good, medium, and bad continuity (as illustrated by Jenyon and Fitch, 1985).

plays examples of seismic continuity. Continuous reflections are characteristic of situations in which depositional environments are laterally similar for considerable distances, as in deepwater sedimentary deposition.

A discontinuous reflector is one that has characteristics similar to those of seismic waveforms across a gap. Such discontinuities may occur because of faults or sudden changes in lateral facies. In the case of faults, we usually see a displacement of continuous reflections. Tools such as the coherency cube (which we will discuss later) are useful for identifying faults. Discontinuities caused by facies changes exist in fluvial or alluvial environments, as found in sedimentary deposition by rivers and streams. In such cases, we generally see a sudden fading out of reflections because of depositional changes.

However, discontinuities may or may not result from geologic reasons. Seismic continuity is generally a function of signal-to-noise ratio for the seismogram. Noise often is caused by acquisition conditions such as wind, wave turbulence in marine recording, or statics. In addition, data processing can introduce noise on the seismogram, and improper stacking velocities, static corrections, or migration can cause disruption of reflection continuity.

Conversely, we also should be aware of artificial continuity caused by processing. A sobering example results from taking a "seismogram" of random noise values. The application of static corrections can align these arrivals to produce apparently continuous seismic reflections. There is a good example of this in Badley (1985, p. 70).

Now that we have some working definitions of continuity, let us look at some quantitative measures of continuity — coherency measures.

# Coherency

Coherency is a measure of the similarity of seismic events. Sheriff (1991, p. 37) offers the following descriptions of coherency. It is defined as "a measure of the similarity of two functions or portions of functions." It is considered "the property of two wave trains having a well-defined phase relationship, i.e., being in-phase." Although many geophysicists may consider coherency qualitatively, quantitative coherency measures are used in automatic picking programs.

Two of the most popular coherency measures used in geophysics are the concepts of crosscorrelation and semblance.

# Crosscorrelation and semblance

Both crosscorrelation and semblance are measures of similarity that one signal shows to another signal. Therefore, they are both coherency measures. Crosscorrelation measures similarity by examining the sum of products of seismic amplitudes. Semblance measures trace similarity by examining the energy of the sum of trace amplitudes compared to the sum of trace energy. Let us examine their mathematical definitions.

Consider a set of traces $x_j(t_i)$ where $j$ denotes trace number and $i$ denotes the time sample. For the case of two traces, $x_1(t_i)$ and $x_2(t_i)$, the normalized croscorrelation is given by

$$r_{12}(\tau) = \frac{\sum_i x_1(t_i)x_2(t_i + \tau)}{\left[\sum_i x_1^2(t_i)\sum_j x_2^2(t_j)\right]^{1/2}}. \tag{3}$$

The numerator of this expression multiplies the amplitude of one signal by a shifted version of another signal and then adds the products of these signals. The denominator is a measure of the energy of the two signals. Note that the numerator is very similar to the convolution formula. In fact, crosscorrelation could be described as a "shift, multiply, add" operation; convolution could be described as a "flip, shift, multiply, add" operation. To illustrate crosscorrelation, let us compute the crosscorrelation of two traces whose digital values are (1,2,0) and (0,1,2).

The energy of both traces equals the sum of the squares of their amplitudes, which in this case is 5. The denominator equals the square root of 25, or 5.

Let us examine the crosscorrelation for various lag values. For the case of $\tau = 0$,

$$r_{12}(0) = \frac{1*0 + 2*1 + 0*2}{\sqrt{(1 + 2^2)(1^2 + 2^2)}} = \frac{2}{5}. \tag{4}$$

For the case of $\tau = 1$,

$$r_{12}(1) = \frac{1*1 + 2*2}{\sqrt{(1^2 + 2^2)(1^2 + 2^2)}} = 1. \tag{5}$$

For the case of $\tau = 2$,

$$r_{12}(2) = \frac{1*2}{\sqrt{(1^2+2^2)(1^2+2^2)}} = \frac{2}{5}. \tag{6}$$

Therefore, we note that the maximum crosscorrelation value is obtained when the second trace is shifted earlier in time by one sample to obtain alignment. This is shown by the fact that a time lag of 1 gives the maximum crosscorrelation. We also note that the maximum value of the crosscorrelation is unity, and this is obtained only when one sequence is a scalar multiple of the other sequence. Therefore, crosscorrelation tells us the similarity of one sequence to another. It tells us if one time sequence has been shifted relative to another and gives us the magnitude of the shift. Consequently, crosscorrelation is used in determining time shifts in static-correction programs.

Another measure of coherence is semblance, a measure that relies on the sum of adjacent traces. If traces are aligned and are similar, the summation of adjacent traces will be greater. The mathematical definition of the semblance of a group of $M$ traces over a time window of $N$ samples is given by

$$S_{MN} = \frac{\sum\limits_{i=1}^{N}\left[\sum\limits_{j=1}^{M} x_j(t_i)\right]^2}{M\sum\limits_{i=1}^{N}\sum\limits_{j=1}^{M} x_j^2(t_i)}. \tag{7}$$

The semblance is the energy of the sum of trace values divided by the sum of the energy of the traces. Its maximum value is also 1.0. In the previous example, we can demonstrate that the maximum semblance value is 1 by shifting the first signal to the right by one sample. In that case, the semblance value is given by

$$S = \frac{(2+2)^2+(1+1)^2}{2[(2^2+1^2)+(2^2+1^2)]} = \frac{20}{20} = 1. \tag{8}$$

It is not always clear whether one should use semblance or crosscorrelation. Crosscorrelation has the advantage of giving the difference in phase lags between traces, and semblance values give coherency measures after phase shifts have been applied. Both methods are used widely in exploration geophysics — in static corrections, in velocity analysis and, most recently, in the coherence cube.

# The coherence cube

In 1995, seismic interpretation saw the advent of a new technology that was built on well-known concepts. The coherence cube, described in a paper by Bahorich and Farmer (1995), produced a new tool for examining seismic discontinuities caused by faults, salt domes, meandering streams, and many other interesting features in petroleum exploration. It might be more accurate to say that algorithms detect faults by revealing a lack of coherency, and so the term "noncoherence cube" might be more appropriate.

The coherence cube and related technologies are valuable interpretation aids for detection of faults and other discontinuities. A testimony to the importance of these techniques was the recognition by SEG of the paper "3D seismic attributes using a semblance-based coherency algorithm," by Marfurt et al. (1998), as the best paper in GEOPHYSICS in 1998. This paper compared the C1 algorithm (based on crosscorrelation) to the C2 algorithm (based on semblance). The C1 algorithm computes the crosscorrelation of traces in the two horizontal directions, normalizes these crosscorrelations with respect to trace energies, and then computes the maximum values for lags in the horizontal directions. The coherency measure is the square root of these maximum values. The C2 algorithm involves a similar algorithm, with the semblance measure replacing crosscorrelation.

Several algorithms have been developed for the coherence cube. There are two families — those that measure coherency (such as crosscorrelation and semblance) and those that detect discontinuities by differencing, such as the algorithm proposed by Luo et al. (1996).

Many recent petroleum-exploration case studies feature applications of coherence-cube technology. One, by Carter and Lines (2001), describes case studies from the Canadian East Coast that involved fault imaging using edge detection and coherency measures on 3D seismic data from Hibernia field. That application used coherence technology on depth-migrated data.

An advantage of applying coherence technology to migrated data is shown in the model example of Figure 2a-e. This simple extensional fault model consists of two stratigraphic layers, between which the 45° sloping fault has a throw of 400 m between depths of 1000 and 1400 m. The fault model was used to create an ideal stacked section in Figure 2b. This stacked section has considerable diffraction energy that obscures the fault position. As shown in Figure 2c, differencing the unmigrated data does not clearly define the correct fault position. Figure 2d shows a reverse-time depth migration of the seismogram in which the position of the fault is defined more clearly. An application of the differencing algorithm clearly outlines the position of the fault, as shown in Figure 2e. In this simple fault example, we note that depth migration produces a much clearer image of the fault

**Figure 2.** Fault detection illustrations (adapted from Carter and Lines, 2001). Differencing results of unmigrated and migrated data of a simple fault model. (a) Simple fault model, (b) unmigrated result, (c) differencing result of unmigrated section, (d) migrated result, and (e) differencing result of migrated section.

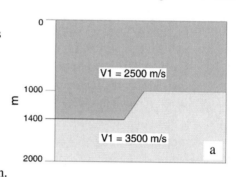

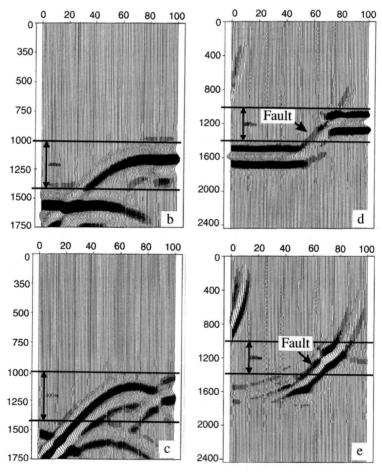

than in the case of unmigrated data, and the subsequent edge detection or differencing algorithm clearly outlines the position of the fault.

The coherence cube showed considerable promise in its ability to enhance interpretation of time slices from 3D data volumes. A time slice is a horizontal slice through a 3D seismic data volume that allows the seismic data to be displayed in map form. In many cases, time slices are difficult to interpret and probably are underused. Faults, for instance, are often difficult to see and interpret from time slices. However, horizontal slices through a coherence cube make faults much easier to detect and map. Bahorich and Farmer (1995) show a nice example of the difference between the interpretability of 3D seismic-data time slices before and after coherency mapping. This comparison (which appeared on the front cover of the October 1995 issue of *The Leading Edge*) is shown in Figure 3. Generally speaking, faults are outlined much more clearly by coherency technology.

## Summary

The concepts of character, continuity, coherence, and correlation are probably intuitively familiar to any interpreter who tries to follow events on seismic sections. This chapter attempts to define these concepts and to place quantitative measures on them. The quantification of coherence has led to algorithms that reside on seismic-interpretation workstations today.

**Figure 3.** Comparison of a time slice (left) and coherency slice (right) (from Bahorich and Farmer, 1995).

# References

Badley, M. E., 1985, Practical seismic interpretation: International Human Resources Development Co.

Bahorich, M. S., and S. L. Farmer, 1995, 3D seismic discontinuity for faults and stratigraphic features: The coherence cube: The Leading Edge, **14**, 1053–1058.

Carter, N., and L. R. Lines, 2001, Fault imaging using edge detection and coherency measures on Hibernia 3-D seismic data: The Leading Edge, **20**, 64–69.

Jenyon, M. K., and A. A. Fitch, 1985, Seismic reflection interpretation: Borntraeger Pub.

Lines, L. R., and S. Treitel, 1985, Wavelets, well logs, and Wiener filters: First Break, **3**, 9–14.

Luo, Y., W. Higgs, and W. Kowalik, 1996, Edge detection and stratigraphic analysis using 3D seismic data: 66th Annual International Meeting, SEG, Expanded Abstracts, 324–331.

Marfurt, K. J., R. L. Kirlin, S. L. Farmer, and M. S. Bahorich, 1998, 3D seismic attributes using a semblance-based coherency algorithm: Geophysics, **63**, 1150–1165.

Sheriff, R. E., 1991, Encyclopedic dictionary of exploration geophysics: SEG.

# Chapter 10

# Modern Solutions to Pitfalls in Seismic Interpretation

One of the most useful books available to exploration geophysicists is the monograph by Tucker and Yorston (1973), *Pitfalls in Seismic Interpretation*. It starts with a biblical quote from Ecclesiastes 10:18: "He that diggeth a pit shall fall into it." This implies that many of the traps found in seismic data are not always of the hydrocarbon variety. These traps are misinterpretations of what we see in the seismic section. They are sometimes of our own creation. Many can be cured by processing.

We will not attempt to recreate all of the excellent examples in Tucker and Yorston (1973). Almost all of the interpretation pitfalls can be summarized into one of three categories:

1)  interpretation of seismic time sections as if they were depth sections, thus failing to recognize velocity effects

2)  interpretation of 3D effects on a 2D seismic section

3)  failure to recognize that some seismic arrivals are not related to the desired geologic structures but may be caused by "noise"

Many of the pitfalls originally described by Tucker and Yorston (1973) can be obviated with modern acquisition and processing methods. For the three pitfall types, modern preventive methods include the following:

1)  If misinterpretation of time sections is a problem, we should convert data to a depth section. This can be done with depth migration. Prestack depth migration is a great tool, as described in later chapters, but it does require accurate velocity information. A good example is shown in Figure 1. From the time section in Figure 1a, it appears that the top of the salt reflector is discontinuous. This is largely the result of velocity effects. Because of high P-wave velocity in the salt (4500 m/s) com-

pared to surrounding sediment (with a P-wave velocity of about 2250 m/s), the section is distorted. If this seismic section is converted to depth by prestack depth migration, the salt reflector appears to be continuous. Actually, there is another pitfall, which is related to 3D effects, as will be shown in Chapter 14 on salt intrusions.

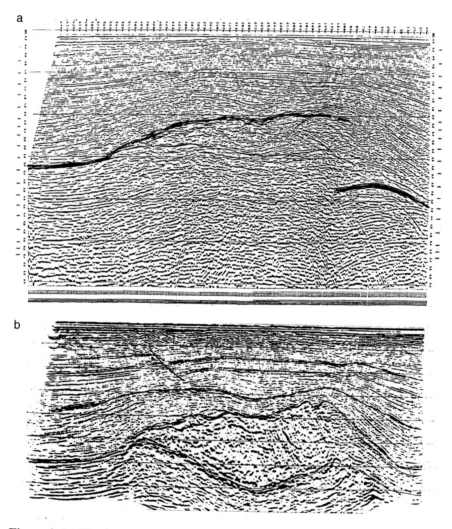

**Figure 1.** (a) Unmigrated seismic section from the Gulf of Mexico in which the top of the salt reflector is marked as a reflector; (b) depth migration of the seismic section in (a). The seismic sections are 19,202 m (63,000 ft) long. The upper seismic section has a total reflection time of 6.1 s. The lower seismic section has an image depth of 4788 m (16,000 ft), as described by Lines (1991).

2) The 3D nature of the subsurface requires 3D acquisition and imaging methods (the topic of Chapter 18 of this book). Modern seismic exploration uses areal arrays of sources and receivers to acquire seismic data in a 3D sense rather than sparsely spaced seismic lines.

3) Many advances have been made in recent years in suppression of "noise" within seismic sections. Modern processing or acquisition methods such as dual sensors can suppress multiples, which are considered organized "noise" or "undesired signal." Nevertheless, interpreters are still stung by interpreting interbed or water-bottom multiples as primary reflection energy.

# Anisotropy — A "modern" pitfall

Although we call anisotropy a "modern" pitfall, it has existed for some time. *Modern* refers to pitfalls that can occur even with the use of routine depth-migration processes. The pitfall that has become most obvious in recent years is that of anisotropy.

Anisotropy is the dependence of a physical property on direction. In seismic anisotropy, this generally refers to the dependence of seismic velocity on the direction of travel.

Seismic waves generally travel faster through most flat-layered sedimentary rocks (especially clastics) horizontally rather than vertically. If we do not account for this anisotropy, the resultant errors in seismic imaging can lead to location errors in drilling a well.

As we see from our analysis of normal moveout (NMO) in Chapter 5, seismic-velocity estimation from NMO often is affected by horizontal-wave velocity. In layered media, horizontal velocities are generally larger than vertical velocities. Therefore, vertical depth conversion using NMO velocities for flat layers may mean that we have overestimated formation depths, although this is not as serious as the anisotropic effects on dipping horizons, which have been pointed out by Lawton (1999) and by Vestrum et al. (1999).

Figure 2 shows the case of dipping shales over a forelimb anticline. Reflections from the crest of the anticline will travel more quickly to receivers on the foreland side in the updip direction.

When these fast traveltimes are imaged using isotropic methods, the imaged position will be shifted too far to the foreland side, as shown in Figure 2. If this image is used for locating a well drilled on the crest of the anti-

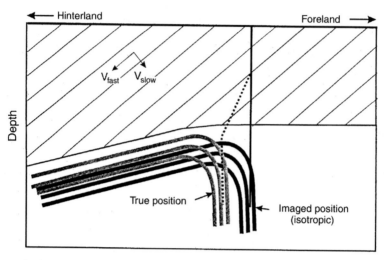

**Figure 2.** The presence of anisotropic dipping shales causes mispositioning of the target unless we account for this anisotropy, as shown by Lawton (1999). The vertical solid line shows the location of a well that would be drilled based on the isotropic image of forelimb folds. The dotted line shows a necessary sidetracked well that would need to be drilled to reach the actual location of the folded beds.

cline, it will miss the target, which can be very expensive. At the very least, the well will need to be sidetracked to hit the structure. This pitfall can be cured by using anisotropic depth migration, which will reverse the reflection along the correct path and produce the true image location.

Figure 3 shows the contrast between isotropic and anisotropic imaging results for an anisotropic physical model. The positioning of the fault is achieved more accurately by anisotropic depth migration. In other words, failure to account for anisotropy can create pitfalls. Accurate anisotropic depth migration is the remedy.

# Conclusions

Pitfalls often arise when we make incorrect assumptions about signals in seismograms or the nature of the geologic model in a seismic image. Most pitfalls can be avoided by the use of depth images, 3D imaging, and good signal-to-noise ratio.

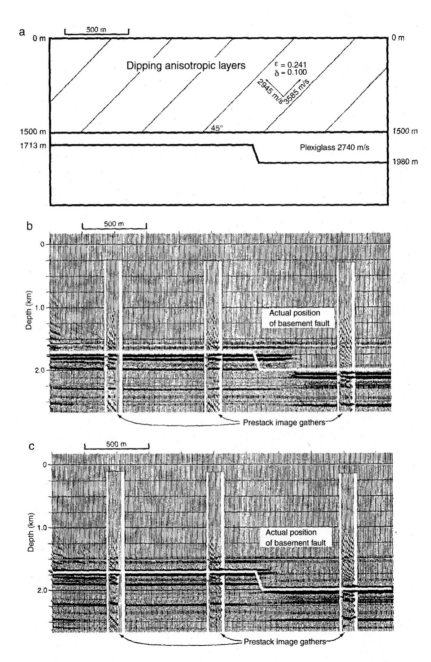

**Figure 3.** Illustration of the effects of anisotropy and the need for anisotropic depth migration (from Vestrum et al., 1999). (a) The physical model, (b) the isotropic migration, and (c) the anisotropic migration.

# References

Lawton, D. C., 1999, Slip-slidin' away — Some practical implications of seismic velocity anisotropy on depth imaging, *in* L. R. Lines, D. C. Lawton, and S. H. Gray, eds., Depth imaging of foothills seismic data: Canadian Society of Exploration Geophysicists.

Lines, L. R., 1991, Applications of tomography to borehole and reflection seismology: The Leading Edge, **7**, 11–17.

Tucker, P., and H. Yorston, 1973, Pitfalls in seismic interpretation: SEG.

Vestrum, R., D. C. Lawton, and R. Schmid, 1999, Imaging structures below dipping TI media: Geophysics, **64**, 1239–1248.

# Chapter 11

# Interpreting a Structurally Complex Seismic Section

## Introduction

The task of the seismic interpreter is to determine the subsurface model that best fits all available geologic and geophysical data. As the areas in which we explore become increasingly complex, we find that the role of interpreters becomes more important and their background must become more diverse. It is essential that interpreters have an understanding of the acquisition and processing of reflection seismic and borehole data. They must be knowledgeable about stratigraphic and structural concepts. With a good appreciation of geology and geophysics, interpreters can provide realistic views of the subsurface and can further our understanding of the earth.

To make a good structural interpretation, an understanding of structural geology and deformational regimes is required. For those who do not have a complete background in thrust tectonics, we recommend the following reference material: Boyer and Elliot (1982), Dahlstrom (1969), and Mitra (1986). For a comprehensive listing of thrust-tectonic terms, see McClay (1992). In addition, we recommend researching the geographic area of interest before starting a structural interpretation so that both the local stratigraphy and deformational styles are understood. For research on the Alberta Foothills, which is the subject of our sample interpretation study, we direct the reader to papers by Bally et al. (1966), Dahlstrom (1970), and Fermor (1999).

In this chapter, we present eight basic steps to seismic interpretation (Table 1) and use a seismic line from the Shaw Basing area of the Alberta Foothills as an example (Figure 1). Each interpretation step is discussed fully in the text. The Shaw Basing line was studied intensively by Yan and Lines (2001) and by Yan (2002), and it presents many interesting interpretation challenges.

# Examine the section

It is useful to have multiple copies of the section and a variety of colored pencils for working through an interpretation. One of the most significant steps is to take enough time examining the section (Figure 1).

**Table 1.** Eight steps to interpretation.

| | |
|---|---|
| 1 | Examine the section. |
| 2 | Add geologic and borehole information. |
| 3 | Interpret large faults and basement formation tops. |
| 4 | Interpret additional formation tops across the section. |
| 5 | Identify and solve interpretation challenges. |
| 6 | Use reflection character where continuity decreases. |
| 7 | Interpret the zone of interest. |
| 8 | Add finishing touches to the section. |

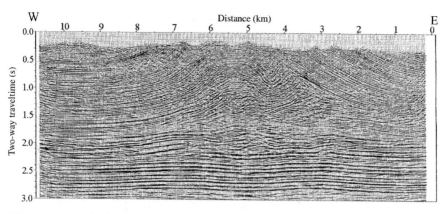

**Figure 1.** Shaw Basing seismic section.

Let your eye follow the structure. Make a note of any significant features or irregularities that catch your gaze. Imagine how the rocks became deformed into that state. Although this may sound strange, it will help give you a feel for the uninterpreted section. Look at the data from different angles. Turn the paper around and make mental notes of where the good data areas are and where the interpretation looks more difficult. Make notes of the locations of continuous reflections across the section. Look for discontinuities, structure, stratigraphy, and geologic features such as reefs or channels. The approach we take is first to interpret the part of the section with good continuity and then use our understanding of structural relationships to interpret the central part of the section, where continuity is diminished and the structure is complex.

In areas of complex deformation, we advise interpreting the simpler parts of the section first. As surrounding reflections are interpreted, the area of interest may become clearer.

## Add geologic and borehole information

Add the additional geologic and geophysical data to the section, i.e., surface geologic information and/or well-log tops. Surface geology is obtained from surface mapping but is not available for this exercise. The tops information, obtained from borehole logs, gives the depth in the borehole of known geologic horizons, typically the top of a formation (Table 2). Draw the trace of the borehole using two-way traveltime (twtt) and distance coordinates given in Table 2. Show well symbols at the top of the borehole to indicate the status of the well (Figure 2). The symbol we show indicates that the boreholes are dry and abandoned. Mark the tops information (Table 2) in varying colors and make note of the color and horizon name in a legend. It is good practice to color in the trough (light area), because the colors are differentiated more easily on a light background than on black. We are constrained here to showing all horizons by black or dashed lines, but you should interpret in color.

It is not always necessary to interpret all the tops given in the borehole. The idea is to have as many horizons as necessary to define the structure of interest. If there are many closely spaced tops, pick a few of the dominant reflectors. Do not clutter the section with dozens of redundant horizons.

In addition, we are not limited to interpreting only the tops from the borehole. Later, it will be advantageous to include more horizons to help define the structure and assist in interpretation. Additional horizons will be named and included in the legend.

In Figure 2, we start our interpretation with the following horizons: Belly River (BR), Wapiabi (W), Bad Heart (BH), Cardium (C), Mannville (M), and Eldon (E). Others will be added as appropriate.

**Table 2.** Borehole information for Well A and Well B.

| Well | Horizontal distance | twtt | Formation top |
|------|---------------------|------|---------------|
| Well A | 9 km | 1350 ms | Belly River |
| Well A | 9 km | 1475 ms | Wapiabi |
| Well A | 9 km | 1600 ms | Bad Heart |
| Well A | 9 km | 1650 ms | Cardium |
| Well A | 9 km | 2000 ms | Mannville |
| Well A | 9 km | 2780 ms | Eldon |
| Well B | 5.85 km | 600 ms | Belly River |
| Well B | 5.80 km | 825 ms | Wapiabi |
| Well B | 5.70 km | 1050 ms | Bad Heart |
| Well B | 5.65 km | 1150 ms | Cardium |

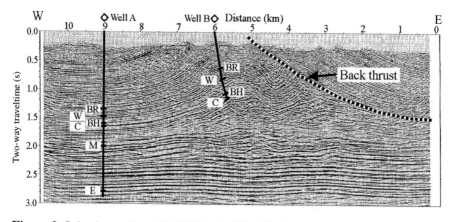

**Figure 2.** Seismic section with Well A and Well B tops. The back thrust is interpreted.

# Interpret large faults and basement formation tops

Now that the horizon tops are marked, take another look at the section. Specifically, look for (1) obvious faults and (2) horizons that are continuous across the section. Note the large fault dipping east that reaches the surface at 5-km offset. Mark this on your section in red pencil. This is referred to as a back thrust (Figure 2). Mark other faults in light pencil so the interpretation can be altered easily.

The easiest horizons to interpret are those that are continuous across the section. The Eldon Formation can be traced across the section, but the Mannville becomes hard to follow beyond the 7-km mark (Figure 3). Mark in the Eldon Formation. It is now useful to mark in the highest reflector that is continuous across the section, with all reflectors below being continuous as well. We have chosen a reflector that intersects Well A at 2.38 s. We arbitrarily named this horizon Top of Basement (TB). The strata below this horizon, hereafter referred to as Basement, are below our area of interest and largely can be ignored.

# Interpret additional formation tops across the section

Now turn your attention to the remaining horizon tops. Starting at Well A, interpret the Belly River, Wapiabi, Bad Heart, Cardium, and Mannville

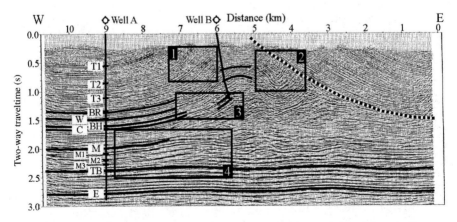

**Figure 3.** Seismic section with horizons partially interpreted. The boxes indicate areas of more difficult interpretation and are shown in more detail in Figures 4–7.

Formations. Always interpret from the flat-lying strata or less complicated area into the more complex area. It is helpful to add new horizons (i.e., T1, T2, T3, M1, M2, and M3) if the quality of the horizon of interest decreases. These additional horizons will help guide interpretation of the primary horizons.

# Identify and solve interpretation challenges

While following the reflectors across the section, there are some hindrances to interpretation, as outlined by Boxes 1–4 (Figure 3). We discuss these boxes separately below.

## Aliasing

**Box 1:** When dips in the section become steeper, the continuity of the reflectors decreases, and interpretation becomes more difficult (Figure 4a). We hypothesize that the decrease in continuity is caused by spatial aliasing, which is discussed in Chapter 7. In the area of steep dip, interpretation becomes more difficult. We are forced to rely on recognizing the character of wavelets or packages of reflectors to correlate from left to right. It is important at this stage to understand the style of deformation. The Alberta Foothills fold-and-thrust belt is dominated by concentric folding (Dahlstrom, 1969). Therefore, there is no thickening or thinning of the beds by flow. Consequently, in the absence of faulting, we will observe constant bed thickness.

Carefully examine the entire domain to make sure that the dip of the reflector matches the surrounding reflectors and that bed thickness is constant. It is very easy to be off by one cycle or leg. In Figure 4b, we illustrate with dotted lines two possible interpretations of the T3 horizon. Sometimes it helps to interpret continuous horizons above and below the reflector of interest (Figure 4c). We observe that the lowermost dashed line (Figure 4b) is more conformable to the dotted horizons above and below the T3; therefore, we interpret the T3 as shown in Figure 4d. The final interpretation (Figure 4d) was made by following continuity, observing the general trend of all the reflections in the domain, and interpreting nearby reflectors with better continuity.

**Box 2:** In this example, crosscutting events affect our ability to interpret data. Although we observe the general trend of the structure along the

Belly River Formation, it is more difficult to interpret T3 and the Wapiabi (Figure 5a). It is difficult to determine which way beds are dipping because continuity is broken by crosscutting events. Figures 5b and 5c show different interpretations for horizons above the Belly River. In Figure 5b, the horizons follow the form of the back thrust. Conversely, Figure 5c shows an interpretation in which the strata are folded upward toward the back thrust. We favor the latter, because the horizons conform to the trend of the Belly River Formation and the thickness of the units is preserved. Our interpretation is based on keeping strata thickness constant (Figure 5d), in keeping with observations documented by Dahlstrom (1969). Although the general

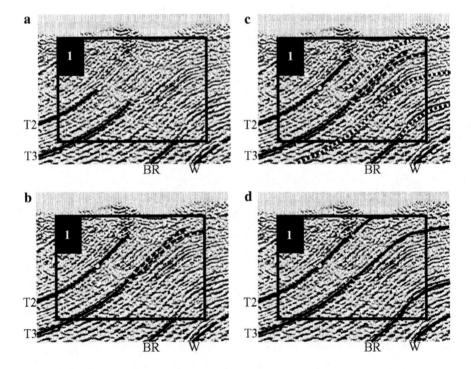

**Figure 4.** Close-up view of seismic section showing area of aliasing. We present a simple method of interpreting aliased data. (a) Part of the interpretation made possible by following the continuous reflections. (b) The dashed lines indicate two possible interpretations for continuation for the T3 horizon. (c) The dotted lines are guidelines along more coherent reflectors used to help determine the location of the T3 horizon. (d) Our final interpretation for the T3 horizon. For the location of Box 1, see Figure 3.

trend of the horizons is well defined, it is difficult to determine where T3 curves upward.

## Well mis-tie

**Box 3:** In this example, we observe a well mis-tie (Figure 6a). There is excellent continuity from Well A until approximately 6.5 km (Figure 2), at which point there is a decrease in continuity. This decrease in data quality does not explain the large mis-tie, but it hinders our ability to recognize

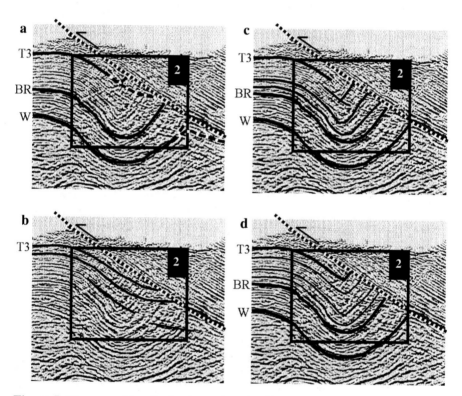

**Figure 5.** Close-up view of seismic section showing an area that contains cross-cutting events. The back thrust is shown with a dotted line. (a) Possible interpretation of the T3, BR, and W horizons. The areas of uncertainty are interpreted with a dashed line. (b) and (c) Two interpretations of the structure between the T3 and W horizons. (d) We prefer this interpretation because the layers retain constant thickness. Dahlstrom (1969) observed that there was no thinning or thickening of beds by flow and that folding was concentric. For the location of Box 2, see Figure 3.

where the horizons terminate. The offsets on the Bad Heart [BH] and Cardium [C] horizons are indicative of faulting, but we are unable to recognize the hanging-wall and footwall cutoffs (Figure 6b).

A cutoff is the intersection of a specific horizon with a fault plane (Dahlstrom, 1970) or, more simply put, the point at which a horizon is cut off by a fault. A hanging-wall cutoff is the fault termination of a horizon that lies in the hanging wall of a fault, that is, above the fault plane. Conversely, a footwall cutoff is the fault termination of a horizon that lies in the footwall of a fault, that is, below the fault plane. The cutoffs must be located to determine the style of faulting.

Both reverse faulting (Figure 6c) and normal faulting (Figure 6d) explain the offset observed on the Bad Heart and Cardium Formations. Strike-slip faulting is discounted because deformation in the Alberta Foothills occurs predominantly within the plane of the seismic section, perpendicular to strike (Dahlstrom, 1969). Because of the lack of definitive reflectors, we must rely on our knowledge of structural style of the area to interpret

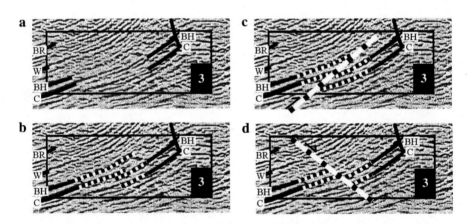

**Figure 6.** Close-up view of seismic section showing a well mis-tie. The Bad Heart and Cardium tops are shown as solid lines to indicate a strong confidence in the interpretation and as dotted lines where confidence in the interpretation is limited. The faults are shown as white dashed lines. (a) Interpretation of BH and C horizons from the borehole. (b) Dotted lines indicate the extension of the horizons and illustrate that there is a mis-tie between the two boreholes. (c) Interpreting the offset between the horizons as a reverse fault. (d) Interpreting the offset between the horizons as a normal fault. For the location of Box 3, see Figure 3.

this offset. The structural regime is compressional; therefore, reverse faulting is favored over normal faulting. With this in mind, we favor the interpretation of Figure 6c but leave the final interpretation of this fault to the finishing touches.

## Fault interpretation

**Box 4:** As the Mannville is interpreted across the section, it becomes unclear which reflector to follow (Figure 7a). The dashed lines indicate two possible interpretations. Note that the package of strata from the Mannville (M) down to the Top of Basement (TB) thickens to the east of Well A, whereas to the west of Well A, the package has constant thickness (Figures 3 and 7a). To help interpret this part of the section, add some horizons between the Mannville and Top of Basement (Figure 7b). Three horizons are shown (M1, M2, and M3).

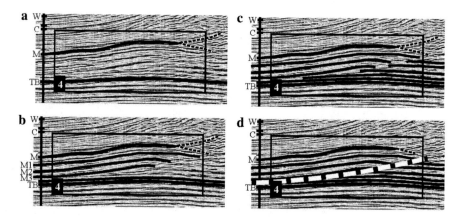

**Figure 7.** Close-up view showing an area of the seismic section in which a fault is required to interpret the reflections. (a) Dotted lines indicate possible interpretations of the Mannville (M) to the east. (b) Additional horizons (M1, M2, and M3) interpreted between the Mannville (M) and Top of Basement (TB). The termination of each horizon is a hanging-wall cutoff. (c) Horizons in the footwall of the fault are interpreted. The terminations of the footwall horizons against the fault are footwall cutoffs. (d) The fault is drawn as a white dashed line. For the location of Box 4, see Figure 3.

Faults rarely are imaged directly on seismic sections. More often, the truncation of reflectors by faults, known as hanging-wall and footwall cutoffs, indirectly delineates faults. Find the hanging-wall cutoffs by following reflectors in the hanging wall until they terminate (Figure 7b). Likewise, trace reflectors from the right-hand side until they terminate in the region of the fault (Figure 7c). Mark the cutoffs with a small dot. The fault now is delineated by small dots and can be drawn on the section (Figure 7d). Although we show the fault as a white dashed line, we recommend interpreting faults using a red pencil. Draw the fault only where it is defined — artistic license is allowed only at the end of the interpretation.

# Use reflection character where continuity decreases

Now that the first interpretation issues are tidied up (Figure 8), look at the upper right part of the section. Earlier, the back thrust was interpreted on the section. The strata above the back thrust are conformable, but we cannot correlate them with our known horizons without well information, surface geology, or character recognition. On a second copy of the seismic section, cut out the portion of section that includes 8.5 to 9.5 km and 0- to 3-s two-way traveltime and mark the trace of Well A and the horizon tops. You need enough section to recognize the character of the reflections.

Overlie the cutout onto the right-hand side of the section and find where the character matches up by sliding the paper up and down the section. If

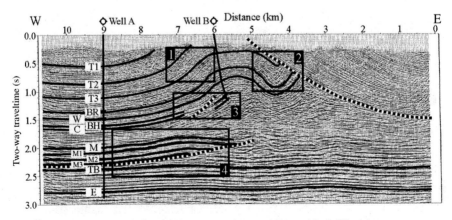

**Figure 8.** Seismic section with interpretation and Boxes 1–4 filled in.

you are unable to find a good correlation, work upward from the Eldon Formation and use approximately the same stratigraphic thickness for each layer. Even when using the latter method, make sure to pick reflectors that have similar character, e.g., do not pick a weak peak for the Mannville on the right-hand side when you marked in a strong trough on the left-hand side. Our picks are shown on the right-hand side of Figure 9.

Interpret T1, T2, T3, BR, and W above the back thrust (Figure 9). Interpret approximately 2 km of the Bad Heart and Cardium Formations, then concentrate on interpreting upward from the relatively flat basement region. This requires correlating across the décollement to locate the Mannville Formation on the east side. We cannot easily correlate across the fault. Therefore, we start to interpret from the right-hand side, where the strata are identified more easily. At that point, you will find it helpful to mark in all the additional horizons between the Mannville and the Eldon.

When you have identified the stratigraphic sequence on the right-hand side, start to interpret toward the west. Note the small fault at 2.75 km at 2 s. A second small structure is observable at 4.75 km on the same horizon. Think about the structural relationships, and interpret these features as you think best. The strata are compressed, and may be interpreted as being either faulted or folded. We interpret the Mannville Formation as a continuous reflector (Figure 10), but this easily may be interpreted as being faulted in parts.

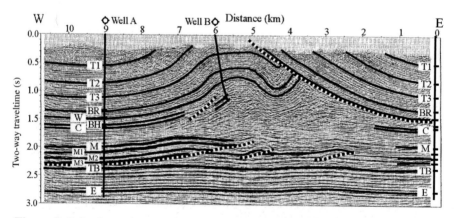

**Figure 9.** Seismic section with borehole information correlated on the eastern part of the section. The solid lines represent horizon tops, and the dotted lines represent faults.

## Triangle-zone core

After interpreting the horizons below the Mannville, it is time to tackle the core of the triangle zone. Figure 10 clearly shows why this structure is called a triangle zone. The flat Mannville horizon is the base of the triangle, and the west-verging back thrust and east-verging fold strata are the two dipping sides of the triangle. The triangle zone is a common structure along the leading edge of deformation in the Alberta Foothills.

As the core of the triangle zone is interpreted, it may be necessary to alter some of your previous interpretation to allow for geologic constraints. We defined the different domains by separating them by faults (Figure 11). Experience tells us that the core of the triangle zone is dominated by reverse faulting because the structure is formed by compression. In a thrust environment, we expect to see repeating layers of strata, duplexes, and multiple thrust faults (Boyer and Elliot, 1992; Mitra, 1986). Although we should not let experience blind us to alternate interpretations, it is reasonable to let experience guide us when the image is not clear.

When interpreting faults, think about basic thrust-fault relationships (Boyer and Elliot, 1982; McClay, 1992). Does the direction of convergence imply that the faulting is normal or reverse? What type of regime is being interpreted? Do any of the faults cut down through the stratigraphic section? Is the interpretation consistent with the geologic understanding of the area?

For the interpretation to make sense, it is necessary to interpret reflectors within the triangle zone. We have already ascertained that the Mann-

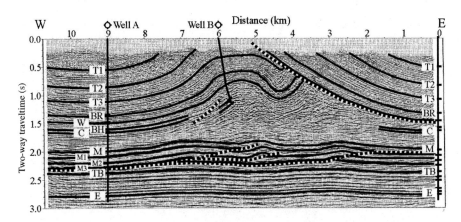

**Figure 10.** Seismic section with pre-Mannville strata interpreted. Horizons are shown as solid lines, and faults are shown as dotted lines.

ville is continuous across the section. The Wapiabi and younger formations are above the main deformation; therefore, we need to be concerned only with the Bad Heart and Cardium Formations. Make sure that faults continue parallel to bedding at depth to observe the structural style of the Rocky Mountains (Dahlstrom, 1969). Put in the reverse fault to correct the well mis-tie.

Figure 12 is our interpretation of the location of reflectors within the core of the triangle zone. We based our interpretation on structural relationships and small portions of reflectors that we interpret as correlatable. The lack of definitive reflectors means that this portion of the section is open to personal interpretation. We recognize that there are places where

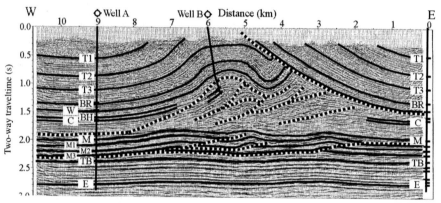

**Figure 11.** Seismic section with faults added to the core of the triangle zone. Faults are shown as dotted lines.

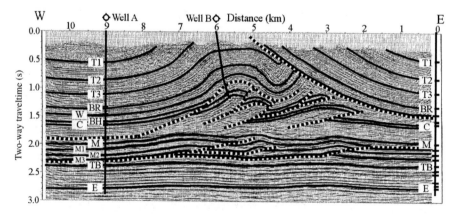

**Figure 12.** Seismic section with Cardium and Bad Heart horizons added to the core of the triangle zone. Faults are shown as dotted lines.

our interpretation could be altered and still observe the structural rules outlined by Dahlstrom (1970). Yan and Lines (2001) present a rather different interpretation of this line based on more recently processed seismic data (Figure 13).

# Add finishing touches to the section

Before the interpretation can be considered finished, faults must be continued down to a décollement and arrows must be added to indicate direction of motion. The final section also will contain a title, name of the interpreter, a legend, and direction markers (i.e., east and west) (Figure 14).

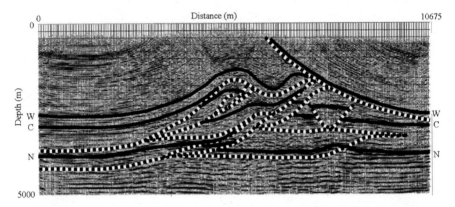

**Figure 13.** Shaw Basing seismic depth section interpreted by Yan and Lines (2001). Horizon tops are labeled on the left-hand and right-hand sides of the seismic section (W = Wapiabi, C = Cardium, N = Nordegg), and dotted lines represent faults.

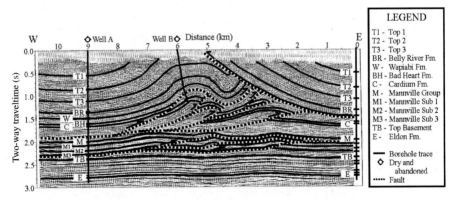

**Figure 14.** Final interpreted seismic section with title, name, date, and legend.

Many interpreters color the entire formation instead of just the formation top. Either practice is acceptable as long as you are not coloring to hide flaws in the interpretation. For this reason, it is always useful to have an uninterpreted section on hand for comparison when looking at an interpretation.

Remember that seismic interpretation is your interpretation of the data. If personal interpretation is not involved, it is just coloring. If your interpretation is consistent with the surficial geologic information, the borehole geologic information, and the principles of structural geology, different interpretations are acceptable and encouraged. We often make a few different interpretations to better assess possible geometries for the underlying structure.

There are many methodologies for interpreting seismic data, of which we have presented only one. As you interpret more data, you will develop your own style. We encourage you to keep the basic principles in mind, regardless of the tectonic regime or your level of experience. As a geophysicist, it is often very easy to forget the geology, even though it is the geology that we want to characterize with seismic interpretation.

# References

Bally, A. W., P. L. Gordy, and G. A. Stewart, 1966, Structure, seismic data and orogenic evolution of southern Canadian Rocky Mountains: Bulletin of Canadian Petroleum Geology, **14**, 337–381.

Boyer, S. E., and D. Elliot, 1982, Thrust systems: AAPG Bulletin, **66**, 1196–1230.

Dahlstrom, C. D. A., 1969, Balanced cross sections: Canadian Journal of Earth Science, **6**, 743–757.

Dahlstrom, C. D. A., 1970, Structural geology in the eastern margin of the Canadian Rocky Mountains: Bulletin of Canadian Petroleum Geology, **18**, 332–406.

Fermor, P., 1999, Aspects of the three-dimensional structures of the Alberta Foothills and Front Ranges: Geological Society of America Bulletin, **111**, 317–346.

McClay, K. R., 1992, Glossary of thrust tectonic terms, *in* K. R. McClay, ed., Thrust tectonics: Chapman and Hall, 419–435.

Mitra, S., 1986, Duplex structures and imbricate thrust systems: Geometry, structural position and hydrocarbon potential: AAPG Bulletin, **77**, 1159–1191.

Yan, L., and L. R. Lines, 2001, Seismic imaging and velocity analysis for an Alberta Foothills seismic survey: Geophysics, **66**, 721–732.

Yan, L., 2002, Seismic imaging and migration velocity analysis of Alberta Foothills structural data sets: Ph.D. dissertation, University of Calgary.

# Chapter 12

# Stratigraphy

## Sequence stratigraphy

Stratigraphy is the science of rock strata (layers of sedimentary rocks), and it deals specifically with the character and attributes of strata. Sequence stratigraphy is the study of genetically related strata bounded by unconformities or their correlative conformities (Sheriff, 1991). Seismic stratigraphy is the study and interpretation of information obtained by seismic-reflection profiling to construct subsurface stratigraphic cross sections (Allaby and Allaby, 1999). Sheriff (1991) describes seismic stratigraphy as methods to determine from seismic evidence the nature and geologic history of sedimentary rocks and their depositional environment. Figure 1 shows an example of the stratigraphic boundaries that are defined by seismic data. Before looking at seismic interpretations, we shall visit sequence stratigraphy.

Sequence stratigraphy refers to sediment deposition controlled by four factors:

1) subsidence of the crust as a result of tectonic and/or isostatic forces

2) eustasy (the rise and fall of sea level)

3) sediment influx from rivers and streams

4) climate, especially as it relates to the development of carbonate reefs in tropical environments

In other words, we could view sequence stratigraphy as an attempt to relate sedimentation to sea levels, tectonics, sediment flow, and climate change.

But what is a sequence? We can think of a sequence as a relatively conformable section between sequence boundaries, which we describe below. A parasequence is a subunit of a sequence (Sheriff, 1991). A sechron is the

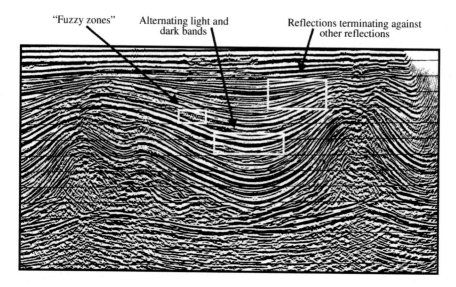

**Figure 1.** Sample seismic data with sequence stratigraphy packets (modified from Yilmaz, 1987, p. 245).

time interval between the top and bottom boundaries. The boundaries and erosional surfaces are determined by the rise and fall of sea level. During a transgression (or relative sea-level rise), deeper-water sediments are deposited, and the transgressive sequence deepens upward. Conversely, during regression, a relative fall in sea level occurs, and the regressive sequence subsequently shallows upward, i.e., shallow-water sediments overlie deepwater sediments. As sea level continues to fall, strata are exposed and erosion occurs, forming a sequence boundary.

Sequence stratigraphy uses an important concept known as Walther's law, which states that the vertical succession of strata represents the horizontal succession of strata. The implication of Walther's law is that our vertical stratigraphic section provides information about the facies that originally were alongside one another. Consequently, when we observe all the lateral facies in our vertical succession, the succession is conformable. The succession is not conformable if lateral facies are omitted. We are interested not only in the presence of an unconformity but also in the type of unconformity.

An *unconformity* is defined as a surface separating younger from older strata, along which there is evidence of subaerial exposure or erosional trun-

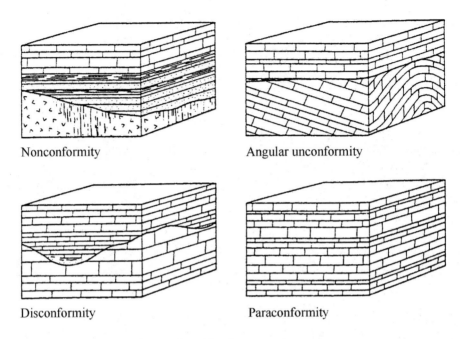

Nonconformity          Angular unconformity

Disconformity          Paraconformity

**Figure 2.** Four types of unconformities (source unknown).

cation, with a significant hiatus indicated (Figure 2). An angular unconformity is the most distinct type: older rocks are tilted and eroded before deposition of younger sediment. We identify the angular unconformity by the angular relationship of the layers. A nonconformity is an unconformity that separates an igneous body from the overlying stratigraphic sequence. A paraconformity is caused by lack of deposition and is characterized by parallel layers above and below the boundary. A disconformity is similar to a paraconformity, except that the older rocks have undergone erosion before the deposition of sediment. The overlying strata are still parallel to the underlying rock, but the boundary will be recognizable as an erosional surface.

The unconformity is identifiable through age-dating techniques, such as examination of the fossil record.

The boundaries may be described further by the relationship between reflections (Figure 3). Onlap is the termination of low-angle reflections against steeper reflections. Downlap is the termination of more steeply dipping reflections against underlying strata that have a shallower dip. If there is uncertainty whether terminations are onlap or downlap, the term *baselap* is sufficient. Toplap is the termination of inclined strata against an overly-

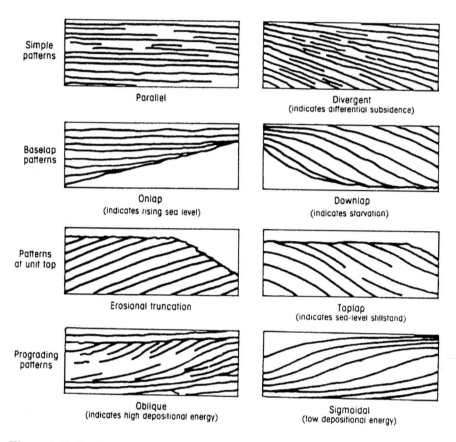

**Figure 3.** Reflection patterns found in seismic sections (Sheriff, 1991; after Sangree and Widmier, 1979).

ing stratigraphic sequence and is usually the result of a period of nondeposition. A slight curvature of the layers into the boundary distinguishes toplap from an angular unconformity, although often it is difficult to differentiate between the two.

Other reflection patterns used in seismic stratigraphy also are shown in Figure 3.

Figure 4 shows a sequence that contains many different reflection terminations and highlights the fact that dipping reflectors may be depositional, not tectonic (i.e., downlap).

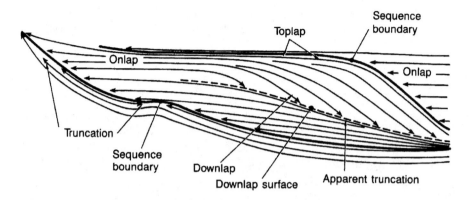

**Figure 4.** Seismic patterns that indicate sea-level changes (Sheriff, 1991, from Vail, 1987).

# Seismic stratigraphy

*Seismic stratigraphy* can be defined as the science of interpreting or modeling stratigraphy, sedimentary facies, and geologic history from seismic-reflection data.

What makes seismic data interpretable? When we look at a seismic section, we see dark bands, bands that terminate, and the general trend of incoherent bands (Figure 1). It is easier to interpret the section geologically by separating it into different components. During this process, we need to remember that the image is not necessarily a true representation of the subsurface. The seismic section may be in time, not depth, and may contain processing artifacts such as multiples.

When we look at a seismic section, we see only the reflections from interfaces, not the layers themselves. With these concepts in mind, Vail et al. (1977) wrote the defining work on seismic stratigraphy, and we recommend it as an excellent reference.

With seismic stratigraphy, we can go beyond a simple interpretation that uses continuity and coherency and can determine the geologic setting. Mullholland (1998) gives an excellent short introduction to sequence stratigraphy. Galloway (1998) discusses clastic depositional systems and sequences with respect to reservoir characterization.

Seismic stratigraphy gives us the knowledge to understand and the terminology to describe seismic reflections with specific geologic interpretations.

# Chronostratigraphy versus lithostratigraphy

Interpretation of seismic data is along isochronous lines, not packages of similar facies. This is illustrated best with a prograding sequence, as shown in Figure 5. As deposition builds up successively seaward, we observe that the facies boundary also migrates. Because facies boundaries are often gradational, we do not commonly observe reflections from them. In contrast, a velocity or density contrast exists between older and younger strata and between distinct facies. Therefore, our reflections represent isochrons.

Seismic stratigraphy can be viewed in chronostratigraphic or lithostratigraphic terms. *Chronostratigraphy* refers to the time of deposition rather than facies surfaces. *Lithostratigraphy* is an element of stratigraphy that deals with lithology (rock type) of the strata. It is interesting that seismic stratigraphy is mostly chronostratigraphic, that is, seismic reflectors are governed by time-depositional units, enabling us to turn back the clock on sediment deposition. Reflection-character analysis allows us to locate where the stratigraphy changes.

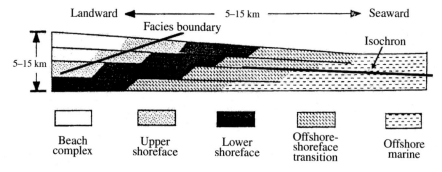

**Figure 5.** Cross-sectional geometry of beds, bedding surface, and sedimentary facies landward to seaward for a prograding shoreline. Note that the facies boundaries are not isochronous surfaces.

# References

Allaby, M., and A. Allaby, eds., 1999, Dictionary of earth sciences: Oxford University Press.

Galloway, W. E., 1998, Clastic depositional systems and sequences: Application to reservoir prediction, delineation and characterization: The Leading Edge, **17**, 173.

Mullholland, J. W., 1998, Sequence stratigraphy: Basic elements, concepts and terminology: The Leading Edge, **17**, 37.

Sangree, J. B., and J. M. Widmier, 1979, Interpretation of depositional facies from seismic data: Geophysics, **44**, 131–160.

Sheriff, R. E., 1991, Encyclopedic dictionary of exploration geophysics: SEG.

Vail, P. R., R. M. Mitchum, and S. Thompson, 1977, Relative changes of sea level from coastal onlap, *in* C. E. Payton, ed., Seismic stratigraphy applications to hydrocarbon exploration: AAPG Memoir 26, 63–81.

Yilmaz, O., 1987, Seismic data processing: SEG.

# Chapter 13

# Carbonate Reef Interpretation

Carbonate reef exploration is extremely important in the Canadian oil industry, with about 60% of petroleum production coming from Devonian reefs. In fact, the advent of modern Canadian oil production originated with the 1947 discovery of the Leduc #1 well. The Leduc discovery marked Alberta's economic advance as Canada's most prosperous province.

It is difficult to imagine that the ancient Devonian climate in Alberta was tropical and that today's oil deposits are largely the result of coral growth in balmy seas. In many ways, carbonate exploration requires that we turn back the geologic clock to reconstruct the setting for the creation of carbonate rocks as traps for oil and gas, because coral reefs require certain conditions of water depth and temperature. It will be our task to image carbonate reefs with seismic exploration tools. A good overview of seismic interpretation of reefs in Canadian exploration is given by Kuhme (1987).

Carbonate reefs generally are considered to be stratigraphic traps, with trapping mechanisms resulting from changes in lithology in a sedimentary environment.

In carbonate exploration, seismic methods usually do not involve imaging of steeply dipping beds. However, advanced migration methods are useful for collapsing seismic diffractions at the edge of the reef.

The analysis of carbonate reefs requires knowledge of the ancient shoreline, as shown in Figure 1. Sedimentary basins usually exist in deeper waters away from the shoreline, creating an environment for offshore shale deposition.

Shallow inland waters on the continental shelf are protected by a barrier reef formed by coral growth. A good example of such barriers is found along the Great Barrier Reef of the Australian shoreline. In the shallow waters of the shelf, reef growth creates mound-shaped patch reefs. Islands of coral growth can create pinnacle or island reefs in the area between the shelf and deeper waters. A diagram of an island reef is shown in Figure 2.

Carbonates are usually of higher seismic velocity than surrounding sediments are. Tight (low-porosity) limestone often has velocities of more than 6000 m/s. In the case of basinal shale, velocities are often in the range of 4000–5000 m/s. Porous carbonates, such as those created by the dolomitization of limestone, have velocities between those of tight limestone and shale. It is this velocity regime that creates many of the seismic features of carbonate reefs.

Sheriff and Geldart (1995) list several seismic characteristics for reef identification. These are summarized below and are shown in Figure 3:

1) Reefs often have a mound shape.

2) Porous reefs often show a reflection void, or dimming. Dimming is caused by a smaller velocity contrast between shale and porous carbonate than between shale and tight limestone.

3) Reef edges often create diffractions that can be observed in unmigrated data.

4) Abrupt terminations of reflections often exist along the flanks of the reef.

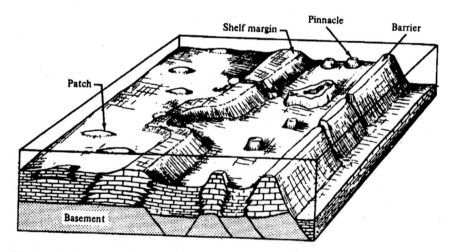

**Figure 1.** Illustration of the environment for reef growth and the types of reefs (from Bubb and Hatlelid, 1977, and in Sheriff and Geldart, 1995).

5) Different sides of the reef have different reflection patterns. One side is generally in a basinal environment; the other side is on the shelf.

6) Layers over the top of the reef have differential compaction, or "drape." The reef represents a high structure overlain by other sediments.

**Figure 2.** Model showing an island reef enclosed by shales. (from Lines, 1991)

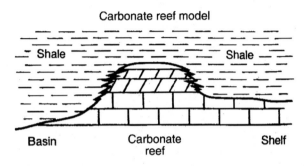

Carbonate reef model

**Figure 3.** The criteria for reef identification, shown by Bubb and Hatlelid (1977) and later by Sheriff and Geldart (1995): (a) mound shaped, (b) reflection dimming, (c) diffractions from edges of reef, (d) termination of reflections, (e) differing reflection patterns on opposite sides of reef, (f) drape over the reef, (g) velocity pull-up beneath reef, (h) velocity pull-down (less common than example g), (i) reef growth on hinge line, and (j) reef growth on structural uplift.

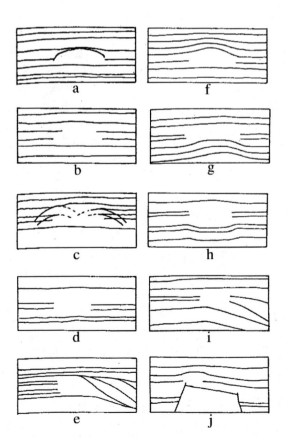

7) Reflectors below the reef usually exhibit a velocity pull-up on a seismic time section, which occurs when subreef reflections arrive earlier in time. This is caused by seismic energy having passed through high-velocity carbonate, rather than through rocks of lower seismic velocity that surround the reef.

8) It is also possible for carbonates to be surrounded by higher-velocity rocks, causing a velocity pull-down, or delay on the time section. This is much less common than the case described in item 7.

9) Reefs often grow on a hinge line or shelf edge.

10) Reefs may grow on a structural uplift caused by faulting. One of the classic examples of reefs is the Horseshoe Atoll in the Permian Basin of west Texas.

Figure 4 shows a seismic stacked section prior to migration. This figure illustrates many of the seismic characteristics of a carbonate reef, as listed above.

The approximate edges of the reef are outlined by arrows. The reef has a mound shape, which is evident in the time window between 1.3 and 1.5 s on the section. There is some drape in reflector 2 over the top of the reef. Reflector 3 shows some dimming over the basinward side of the reef — probably a result of porosity development caused by dolomitization.

Beneath this mound-shaped structure is a velocity pull-up of reflectors 3 and 4. This apparent high, or pull-up structure, on the time section is caused by the early arrival of reflected seismic waves passing through high-velocity limestone, as compared to reflected arrivals that travel through low-velocity basinal shales. The reflectors are fairly flat in depth, but the velocity effects cause an apparent high feature on the time section. Reflections on the basinward (left) side of the section have a different character than those on the shelf (right) side.

In the section shown in Figure 4, there are also diffractions near the edges of the mound. These diffractions subsequently are collapsed in the depth migrations of Figure 5.

Figure 5 shows the superposition of a velocity tomogram over the depth-migrated section. A velocity tomogram is a matrix of interval velocity cells derived from traveltime tomography (as described in Chapter 17). The basinal shale between depths of 7000 and 8000 ft and between $x = 0$ and $x = 15,600$ ft has a lower velocity than the high-velocity carbonate at the same depths for $x > 15,600$ ft. Note that the depth migration removes the

Events                                                    Time (s)

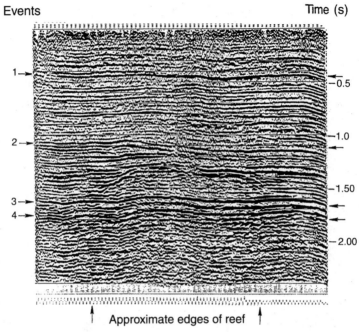

Approximate edges of reef

**Figure 4.** Seismic example of the Horseshoe Atoll, presented at the 1991 SEG Spring Distinguished Lecture Series by Lines (1991). Horizontal distance is 15,800 m (52,000 ft).

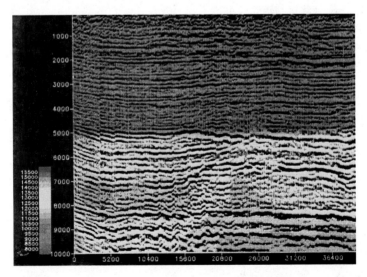

**Figure 5.** Depth migration and velocity tomogram overlay for the reef and off-reef portions of the seismic data in Figure 4 (from Lines, 1991). The vertical and horizontal scales are in feet and the velocities are in feet per second. The section covers the western 12,192 m (40,000 ft) of the area shown in Figure 4.

apparent pull-up of reflectors 3 and 4, which was evident on the time section of Figure 4. (The reflectors beneath the reef actually are fairly flat in the depth domain.)

For reef interpretations, there are advantages to looking at seismic sections in both the time and depth domains. Velocity is the key to linking those domains. Both velocity and seismic reflectivity show differences among basinal shales, porous carbonates, and tight carbonates. For island reefs and shelf carbonates, porosity increases are indicated by both velocity decrease and reflector dimming. It will be more difficult to determine whether the pores are filled with gas, oil, or brine. For different fluids, the seismic responses show subtle differences. Seismic interpretation of reef traps remains an important and nontrivial exercise.

# References

Bubb, J. N., and W. G. Hatelid, 1977, Seismic recognition of carbonate buildups in seismic stratigraphy, *in* C. E. Payton, ed., Seismic stratigraphy — Applications to hydrocarbon exploration: AAPG Memoir 26, 185–204.

Kuhme, A. K., 1987, Seismic interpretation of reefs: The Leading Edge, **6**, 60–65.

Lines, L. R., 1991, Applications of tomography to borehole and reflection seismology: The Leading Edge, **7**, 11–17.

Sheriff, R. E., and L. Geldart, 1995, Exploration seismology: Cambridge University Press.

# Chapter 14

# Interpretation of Traps Related to Salt Structures

## Overview

In the last decade, many of the important oil-field discoveries have been a result of exploration of Gulf of Mexico salt traps. Possible salt-related petroleum traps also exist in onshore Texas and Louisiana, the Paradox Basin of Utah, the North Sea, North Africa, the Hannover Basin of Germany, and the Caspian Sea area. Other areas such as the east coast of Canada have potential petroleum traps associated with salt intrusions but have yet to provide significant production.

Salt is unusual in that it does not follow the normal seismic velocity–density relationships of many other rocks. Salt generally has a lower density and often higher seismic velocity than surrounding sediments. With the low density and "buoyancy" of salt buried under sediments, salt usually will "float through," or intrude through denser sediments, and it often will pierce overlying sediments, providing traps along flanks.

Various salt shapes can result from these intrusions. Salt (halite, or NaCl) often provides a permeability barrier for petroleum in sediments that dip against the salt flank. In addition to this type of trap, significant faulting often is associated with intrusion. Sealing faults can serve as traps above salt intrusions. In other cases, petroleum may accumulate beneath intrusions.

Levorsen (1967) classified salt domes according to their age. Young domes are characterized by anticlines with relatively little deformation. Mature domes are vertical salt stocks on which cap rock has begun to accumulate. Older domes have thicker accumulations of cap rock that is a disk-like body of anhydrite, gypsum, and carbonate overlying the salt body.

Seismic prospecting in areas with salt-intrusion traps has produced some very challenging imaging problems. Because of the 3D shape of salt, there is generally a need for 3D migration. In Gulf Coast sediments, salt often has a much higher P-wave velocity than surrounding rocks, and imaging problems arise around this large, lateral-velocity contrast between the salt and surrounding sediments. For example, in the Gulf Coast area, the velocity of salt is about 4500 m/s, and surrounding sediments have much lower velocities, of about 2250 m/s, yielding a difference of a factor of two in interval velocity.

Time-migration methods normally assume there are no lateral-velocity variations. Therefore, to image Gulf Coast salt bodies successfully, one must use depth-migration methods that do not make the restrictive assumption of a constant lateral velocity. In addition to the velocity contrast, salt bodies intruding through sediments often have very steep dips (many have vertical or overturned flanks) and rugose surfaces that would violate the use of poststack processing assumptions. The lateral-velocity variation, 3D salt shape, and steep dips have made it desirable to use 3D prestack depth-migration methods — the most expensive of imaging techniques. Generally, a sequence of migration steps and velocity-analysis steps (with interpretation at each stage) is used to provide useful images of possible petroleum traps. Fortunately, computer speed, large memories, inexpensive disk space, and parallel computing now make 3D prestack depth migration feasible, tractable, and affordable.

Ratcliff et al. (1992) and Ratcliff et al. (1994) elegantly demonstrated the application of sequential migration processes that can successfully image salt intrusions. These papers, published in *The Leading Edge*, are case histories that deal with 3D depth imaging of structures in the Gulf of Mexico.

Ratcliff et al. (1994) compare the use of migration methods, starting with simple 2D poststack time-migration methods and progressing to 3D prestack depth-migration methods. It is interesting to see the evolution of velocity models and images that start with simple, cheap migration methods and progress to general, expensive migration methods. It is sobering to note that it is perilous to ignore 3D effects when exploring salt traps.

This is well illustrated in the *TLE* paper by Ratcliff et al. (1994), dealing with subsalt exploration of the Vermilion structure in the Gulf of Mexico. The paper is included below, largely verbatim, with the permission of the first author and publisher, for the benefit of processors and interpreters involved in petroleum exploration in the vicinity of salt intrusions.

# Subsalt imaging via target-oriented 3D prestack depth migration

*by D. Ratcliff, C. A. Jacewitz, and S. H. Gray*

We discuss subsalt imaging by way of target-oriented 3D prestack depth migration. First, we review the need for 3D prestack depth migration to image below the salt. Second, we describe the 3D poststack depth-migration sequence leading up to the 3D prestack migration, as well as the velocity iterations associated with depth migration. Third, we show several comparisons involving 3D prestack depth-migrated data.

We conclude with a look at two depth-migrated seismic lines recorded over a salt sill where we have drilled through salt and have knowledge of subsalt reflectivity as well as subsalt structure. Before we do that, we describe why this technology is so important to us. Through the application of 3D prestack depth migration, we can obtain images such as Figure 1, in which an anticline is clearly visible below a salt body. Imaging results such as this can be used by the explorationist to identify and map subsalt structures that could trap significant hydrocarbon accumulations. The technology

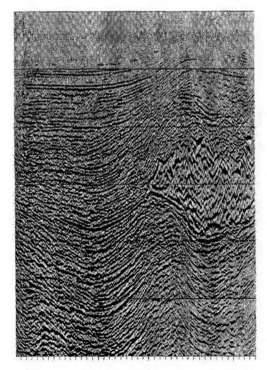

**Figure 1.** 3D prestack depth-migrated subsalt data, 112-fold (from Ratcliff et al., 1994).

also can be used to reduce the risks associated with subsalt drilling and help optimize well locations. So with this in mind, let us proceed to our first topic and review the need for 3D prestack depth migration.

## Need for 3D prestack depth migration

The industry is aware that 3D technology is a lot better than 2D technology. The comparison in Figure 2 shows this. The 2D poststack time migration in Figure 2a provides a picture of the top of salt (TOS), which is highly distorted by the presence of out-of-plane reflection events. The 3D poststack time migration in Figure 2b shows a clear image of the TOS; all TOS reflection energy has been migrated to its correct location in 3D. To image below salt, we first need to get an accurate picture of the top of salt, and 3D poststack depth migration allows us to do this.

The next figure pair shows why we need prestack depth migration. Figure 3a shows a seismic line over a salt sill, processed using 2D prestack time migration. Figure 3b shows the same line processed using 2D prestack depth migration. Notice that the depth-migrated data give us a picture of the subsalt anticline, but the time migration shows a garbled picture below salt. In this area, prestack depth migration is required to obtain subsalt images. So if we combine the benefits of 3D with the benefits of prestack depth migration, we should obtain the very best subsalt images. Indeed, that will become clear.

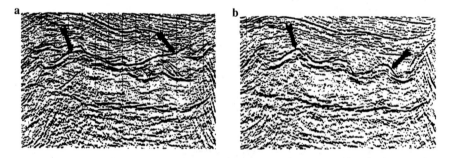

**Figure 2.** (a) 2D poststack time migration showing an incomplete and distorted TOS image (arrows). (b) 3D poststack time migration. 3D migration has removed the distortion (from Ratcliff et al., 1994).

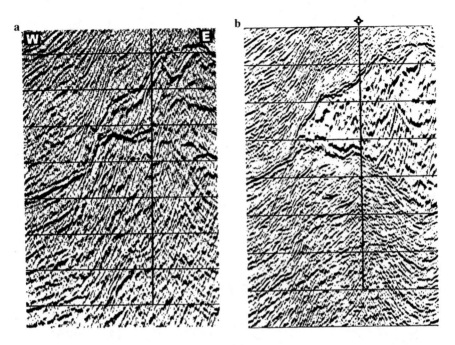

**Figure 3.** (a) 2D prestack time migration showing incorrectly positioned bottom of salt (BOS) and a lack of subsalt reflectors. (b) 2D prestack depth migration. BOS and subsalt images are considerably improved (from Ratcliff et al., 1994).

## 3D poststack depth-migration sequence

We continue to the 3D poststack depth-migration sequence by explaining the velocity iterations associated with these migrations. We use 3D poststack depth migration as a tool to help build the 3D velocity field over the salt structure of interest. This procedure is, of course, much less expensive than iterating on the 3D prestack migrations, but it can be as effective in deriving a 3D velocity field as full 3D prestack depth migration–velocity analysis.

Figure 4 shows the small 15- × 5-km rectangle that covers the area where we performed 3D poststack depth migration. It contains about six million prestack traces. Notice the outline of the TOS and the well control (large dot). To describe the 3D poststack depth-migration sequence, we show depth-migration results from an east-west line indicated by the long arrow. (Although the poststack migrations had the large rectangle in Figure 4 as both input and output data volumes, we show the results on a single line.)

Before the first 3D poststack depth migration, we generate a 3D sediment-velocity field, using a few velocity-model tools. Then, as illustrated by the steps in Table 1, we refine this field into one suitable for correctly imaging subsalt targets (Figure 5). Figure 5a shows a velocity cross section from this interval velocity field; the velocity varies slowly across the line. We next depth-migrate our poststack data using the 3D velocity field to find the correct location of the TOS. Figure 5b shows the east-west seismic line from the survey after 3D poststack depth migration. The first iteration of the depth migration has correctly imaged the sediments above salt (including faults) and the TOS, but it has incorrectly imaged the bottom of salt (BOS) and any subsalt reflections.

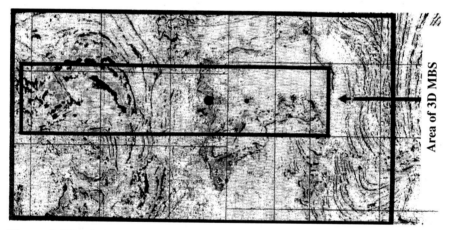

**Figure 4.** 3D-migrated time slice over a salt structure in the Gulf of Mexico (east-west inline through well indicated by arrow). MBS = migration before stack (from Ratcliff et al., 1994).

**Table 1.** Ingredients for 3D velocity model building.

| Sediment velocity field | Well control |
|---|---|
| | 3D DMO velocity field |
| | 2D MBS velocity analysis |
| | 3D poststack depth migration |
| **3D salt and sediment velocity field** | **3D poststack depth migration** |
| | GOCAD 3D velocity package |

From Ratcliff et al., 1994.

At this point, we interpret the TOS in 3D from the first iteration and update our 3D sediment velocity field with the 3D TOS interpretation, as shown in Figure 5c. Notice the TOS interpretation on the velocity cross section. We then use this updated velocity model to remigrate the 3D data, with the purpose of finding the correct BOS location. This migration is shown in Figure 5d. As intended, the second migration has focused the sediments above salt (as before), the TOS, and the BOS, but it has incorrectly migrated the subsalt reflectors because it used salt velocities below salt. Now we interpret the BOS in 3D from the second iteration and use this information to update our 3D velocity field once again. Figure 5e illustrates this update.

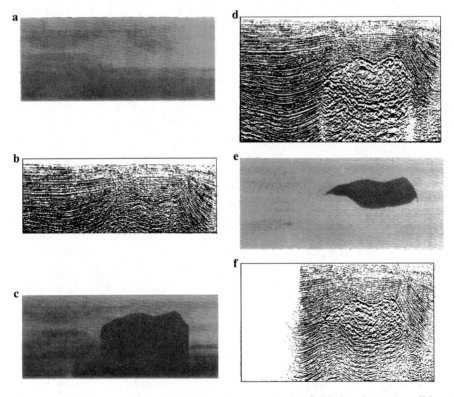

**Figure 5.** (a) Cross section of the 3D sediment-velocity field showing very mild lateral-velocity variations. (b) First iteration (correctly imaged down to TOS) of 3D poststack depth migration. (c) 3D velocity field with TOS update. (d) Second iteration (correctly imaged down to BOS) of 3D poststack depth migration. (e) 3D velocity field with BOS update. (f) Third iteration (correctly imaged beneath salt for those events that have survived stack) of 3D poststack depth migration (from Ratcliff et al., 1994).

At this point, we have the 3D salt- and sediment-velocity field, which we use to perform our third iteration of 3D poststack depth migration to image any subsalt reflection events that may have survived the stacking process.

The final poststack migration is shown in Figure 5f. Now we have the TOS imaged, the BOS imaged, and limited imaging of subsalt reflectors. Although the imaging quality has improved with each iteration of 3D poststack migration, the subsalt image cannot be interpreted with confidence, and we expect the 3D prestack depth migration to improve it.

Table 1 summarizes the procedures and tools used to build the sequence of 3D velocity models. To create the 3D sediment-velocity field, we used a combination of well information, 3D DMO (dip moveout) velocity information, 2D prestack migration-velocity analysis, and 3D poststack depth migration. To create the 3D salt- and sediment-velocity field, we used 3D poststack depth migration, along with the 3D design software GOCAD velocity model–building package.

Now that we have (we hope) established a good 3D velocity field, we can move on to the 3D prestack depth migration. Table 2 shows the processing flow that was used to perform this migration. Initially, we resampled and edited the field data associated with our target, located within the small rectangle of Figure 4. We then gained the data for spherical divergence corrections and performed muting, deconvolution, and filtering. Next, we merged the navigation data with the seismic data and finally performed a 3D common-offset sort.

The output of this sort consisted of many 3D common-offset data volumes. In this survey, there were 120 channels along the cable, so we sorted our data into 120 common-offset data volumes. This 3D sorting step provided many benefits to

**Table 2.** 3D prestack depth-migration flowchart.

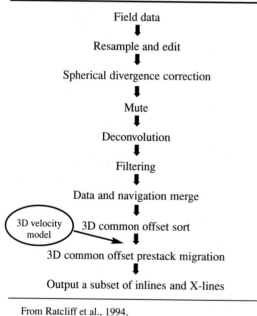

Field data
↓
Resample and edit
↓
Spherical divergence correction
↓
Mute
↓
Deconvolution
↓
Filtering
↓
Data and navigation merge
↓
3D velocity model → 3D common offset sort
↓
3D common offset prestack migration
↓
Output a subset of inlines and X-lines

From Ratcliff et al., 1994.

3D prestack migration, as we will see in the next section. At this point, our data were prepared for 3D prestack migration. Using the final velocity field from our 3D poststack migration iterations, we performed Kirchhoff depth migration on each of our common-offset volumes. Next, we present some comparisons involving the 3D prestack migrated data from the east-west line indicated in Figure 4.

## Comparisons

We consider fold and offset differences first. Then we compare 2D pre-stack with 3D prestack migration results, and finally we compare 3D post-stack imaging with 3D prestack imaging. For Figure 6a, one common-offset data volume was 3D prestack depth-migrated onto a single east-west line, resulting in a singlefold data set. Figure 6b shows the same line with nine common-offset 3D data volumes migrated onto one line (a ninefold stack). Note the subsalt reflections evident on both images, but note also the signal-to-noise (S/N) improvement on the higher fold migration.

S/N continues to improve as we increase the fold to 47 in Figure 6c. We can monitor this S/N improvement with higher fold as we accumulate more common-offset volumes into the stack of migrations. This capability is a major benefit of 3D common-offset migration; it gives us the option to choose which offsets to migrate. Choosing more offsets (including far offsets) permits a detailed migration-velocity analysis, and choosing fewer offsets permits an economical final stack of very high quality. For example, comparison of Figure 6c with Figure 6d shows little interpretable difference between 47-fold and 112-fold. Another benefit of sorting to common-offset volumes is turnaround; several common-offset migrations can be run over-night on a Cray 2 supercomputer to yield fully migrated, interpretable target volumes.

Now we examine an offset comparison. Figure 7a shows our seismic line migrated using offset data volumes ranging from 1300 to 2000 m. Figure 7b shows the same line, migrated with offsets ranging from 375 to 2000 m, and an improved TOS image. The near-offset information contributes more to the TOS image than does the far-offset information. We advance three possible reasons for this effect: amplitude variation with offset (AVO), shooting direction, and salt geometry. Of these, we shall discuss only AVO.

Figure 8, a synthetic common-depth-point (CDP) gather, shows some range-dependent plane-wave reflection responses of the TOS, with the sediment-velocity function that was used in the migration. From the scale on the right of the figure, we can see that the bulk of the reflection energy from TOS comes from the near offsets, which have contributed to Figure 7b but not to Figure 7a. With this in mind, we again find the separate migration of

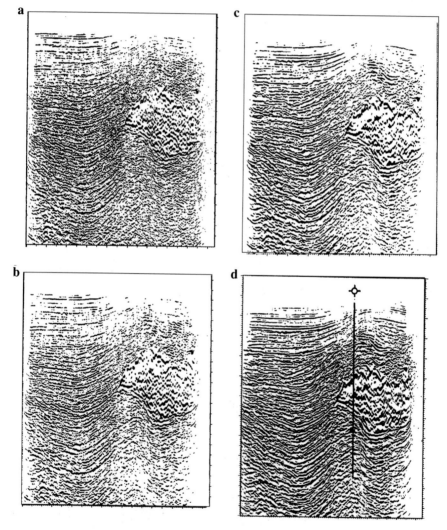

**Figure 6.** 3D prestack depth migration, (a) singlefold (1325 m), (b) ninefold (1275–1400 m), (c) 47-fold (900–2100 m), and (d) 112-fold (from Ratcliff et al., 1994).

common-offset volumes advantageous, because it allows us to choose those offsets that contribute most coherently to the final image.

Figure 9 continues the offset comparison, with the major difference evident beneath salt. Offsets range from 375 to 1600 m in Figure 9a and from 375 to 3000 m in Figure 9b. The fold is 50 in each figure. Differences are subtle, but including a wider range of offsets has enhanced the subsalt

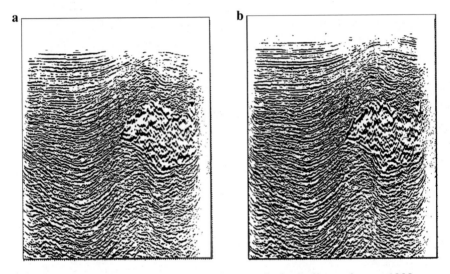

**Figure 7.** 3D prestack migration with (a) range-limited offset volumes, 1300–2000 m, and (b) range-limited offset volumes, 375–2000 m. Inclusion of near offsets has improved the image of TOS (from Ratcliff et al., 1994).

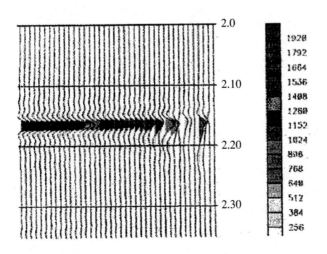

**Figure 8.** AVO synthetic at sediment-TOS interface. TOS response is from all offset ranges; amplitudes decrease as offset increases to the right (from Ratcliff et al., 1994).

image in Figure 9b. We speculate that ray-bending effects account for most of the differences in the subsalt images.

Next, we compare the difference between 2D prestack depth migration (Figure 10) and 3D prestack depth migration (Figure 6d) on the same line. The 2D migration shows out-of-plane TOS and voids in the BOS, whereas the 3D migration has correctly imaged both the TOS and the BOS. Note the clearly imaged subsalt anticlines penetrated by the well.

The final comparison shows the differences between 3D prestack (Figure 6d) and 3D poststack depth migration (Figure 11). The subsalt anticlinal reflectors appear to be crisper and more continuous on the prestack migration.

How correct was our velocity model? Figure 6d shows final 3D prestack depth-migrated data on our east-west line, with the well indicated. Figure 12 shows the 3D-migrated CDP gathers. Given the large number of offset volumes that contributed to these gathers, we consider Figure 12 to be as important as the final stack. The flatness of almost all the events on the migrated CDP gathers increases both the processor's confidence in the

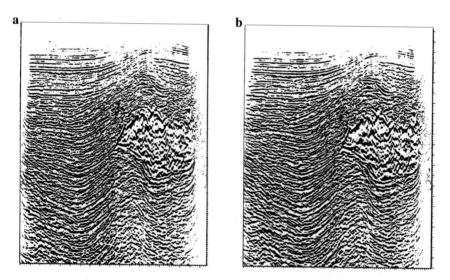

**Figure 9.** 3D prestack depth migration, 50-fold, with range-limited offset volumes, comparing the effect of offset range on subsalt imaging. Offset (a) 375–1600 m and (b) 375–3000 m (from Ratcliff et al., 1994).

migration-velocity field and the interpreter's confidence in the mapped structure.

Now we reach the bottom line, the well results. The subsalt well, drilled in 1988, encountered sands that appear as the anticlinal reflectors on the 3D prestack migrated images. However, these sands did not contain hydrocarbons. This is puzzling, given that the well appears to be drilled on a structural high on our east-west line (Figure 6d). However, a look at the north-south line through the well clears up the puzzle. The 3D prestack depth-migrated data in the north-south direction (Figure 13) show the absence of a structural trap. This well was drilled without the benefits of 3D data. Today, we try not to drill subsalt wells without having used 3D. In fact, we routinely use target-oriented 3D prestack depth migration on our subsalt prospects to avoid drilling subsalt structures that lack closure.

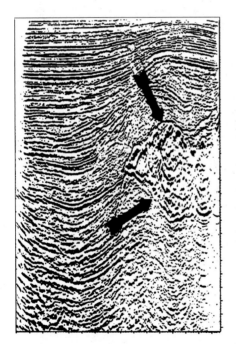

**Figure 10.** 2D prestack depth migration showing out-of-plane TOS and void in BOS (courtesy Frank Sheppard; from Ratcliff et al., 1994).

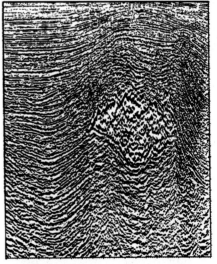

**Figure 11.** 3D poststack migration. The subsalt structure is not as well imaged as in Figure 6d (from Ratcliff et al., 1994).

**Figure 12.** CDP gathers from 3D prestack depth migration. Nearly all the events on the gathers are flat, indicating a correct migration-velocity field (from Ratcliff et al., 1994).

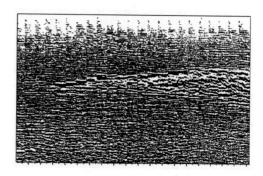

**Figure 13.** North-south line after 3D prestack depth migration with well location indicated. The subsalt structure shows no apparent closure in the north-south direction (from Ratcliff et al., 1994).

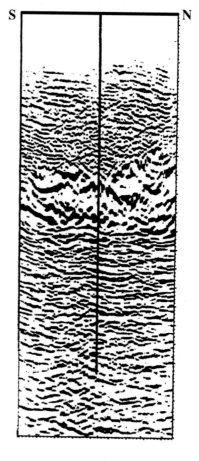

## Summary and discussion

On a subsalt structural play, we have shown that 3D prestack depth migration has provided significantly better imaging than either 2D prestack depth migration or 3D poststack depth migration. Along the way, the migration velocities for the 3D prestack migration were determined systematically, using relatively simple tools. (It is interesting to note that just a few years ago, very few people would refer to iterative 3D poststack depth migration as "simple.") We also have shown that sorting the data into common-offset volumes before migration has several benefits, chief among them flexibility and economy. In fact, we believe that the combination of these benefits takes target-oriented 3D prestack depth migration out of the realm of "grand-challenge" problems and into the realm of present-day technology.

We see this technology helping to open up a new exploration basin — the subsalt Gulf of Mexico. Recent subsalt discoveries have proved that there are significant hydrocarbon accumulations beneath salt. Until recently, the structures holding these deposits had remained all but invisible on seismic records. However, as shown here, the combination of accurate velocity-analysis techniques and powerful prestack-imaging methods can remove a great deal of distortion from a previously blurred picture.

Problems remain, of course, such as pervasive multiple-reflection energy, which continues to obscure our view of subsalt reflectors and seismic amplitudes that have been altered drastically by wave propagation through irregular salt bodies. However, these problems do not make imaging impossible. As subsalt discoveries increase (partly as a result of 3D prestack imaging), the economic demands of developing new fields will drive our industry toward solving these subsalt quality problems, further enhancing our ability to image subtle targets in complicated geology.

## Acknowledgments

Teams in Amoco Exploration and Amoco Production Research interacted closely on this project. We want to emphasize that neither of these teams was capable of delivering these results on its own. Among others, thanks go to Frank Sheppard, Valerie Caspary, John Pritchett, Bill House, Deedee Ragusa, Ron Luongo, Alan Brown, Rod Van Koughnet, and Harley Bandy in Amoco Exploration and to Jo Lynch, John Etgen, and Ken Kelly in Amoco Production Research. Field data are courtesy of Schlumberger–Geco Prakla.

**Note:** The authors of this book thank Davis Ratcliff, Chester Jacewitz, and Sam Gray for allowing us to reprint the excellent article above.

## References

Levorsen, A. I., 1967, Geology of petroleum: W. H. Freeman and Company.

Ratcliff, D., S. H. Gray, and N. D. Whitmore, 1992, Seismic imaging of salt structures in the Gulf of Mexico: The Leading Edge, **11**, 15–31.

Ratcliff, D., C. A. Jacewitz, and S. H. Gray, 1994, Subsalt imaging via target-oriented 3D prestack depth migration: The Leading Edge, **13**, no. 3, 163–170.

# Chapter 15

# Seismic Modeling

## Overview

In describing seismic modeling, it is helpful to review the definition for *model* in Sheriff's dictionary (1991, p. 196): "Model: 1. A concept from which one can deduce the effects for comparison to observations: used to develop a better understanding of the observations. The 'model' may be conceptual, physical, or mathematical."

## Seismic-modeling methods

Seismic modeling attempts to simulate the subsurface rock properties of the earth and the response of seismic-wave propagation as the waves travel through the earth. Seismic models can vary in one dimension (1D), two dimensions (2D), or three dimensions (3D). The accuracy of these models for a real situation depends entirely on the geologic setting. Normally, the model choice is a trade-off between cost and model validity. Although 1D models are less expensive and can be appropriate for flat-lying prairie geology, they would be entirely inappropriate for foothills or salt-dome models, where one would have to rely on 2D and 3D models.

In addition to dimensionality considerations, the cost of modeling will depend on other types of approximations. For example, wave-equation methods are more expensive but can offer more general and complete seismic models than other methods. Seismic-ray tracing is a high-frequency approximation to the wave equation but is usually less expensive than finite-difference or finite-element approximations to the wave equation. As a prelude to our discussions on modeling, we examine the modeling methods in Table 1.

**Table 1.** Modeling methods.

| Modeling type | Mathematical model | Generality | Expense |
|---|---|---|---|
| **Normal-incidence reflectivity** | One-dimensional with reflectivity values given by: $$R = \frac{\rho_2 v_2 - \rho_1 v_1}{\rho_2 v_2 + \rho_1 v_1}.$$ | This is valid only for flat layers and vertically traveling waves. Multiples can be included. | This is very inexpensive for reflectivity and only slightly more expensive if multiples are included. |
| **Amplitude variation with offset (or angle)** | "1.5D" earth model is 1D and offset is nonzero (2D). It uses Zoeppritz's equations. | Strictly speaking, this is valid for a flat-layered earth, and multiples are not generally included. | AVO models are more expensive than normal-incidence reflectivity but less expensive than most wave-equation solvers. |
| **Ray tracing** | 2D, 3D solutions to Snell's law. Amplitudes can be included through asymptotic ray tracing. | This is generally applicable when the scale of the heterogeneity is large compared to the Fresnel zone. It usually ignores diffractions. | Medium expense in most cases. Ray-paths are often cheap to compute, but amplitude calculations can add to the expense. |
| **Wave-equation finite-difference (FD) or finite-element (FE) solutions** | 1D, 2D, 3D numerical solutions to $$\nabla^2 u = \frac{1}{v^2}\frac{\partial^2 u}{\partial t^2}$$ | FD is as general as rectangular grids allow. FE is even more general and uses gridding algorithms. | Expensive — this is the price of generality. |
| **Physical modeling** | 1D, 2D, 3D; requires correct physical materials in scaled models | This model is as general as the physical materials allow. | It is expensive to set up the model itself, but modeling runs are often not as expensive as numerical modeling. |

Discussions of these modeling methods are found in many books, including those by Aki and Richards (2002) and by Kelly and Marfurt (1990).

# Uses of seismic modeling

From the previous discussion, we see that there are many methods for seismic modeling. Likewise, seismic modeling has many uses, and here we list six of them.

## Design of seismic experiments

In constructing a seismic experiment, we are faced with the questions of designing source-receiver geometry. We need to know the number of sources and receivers and the spacing between them. Although this information may be developed from previous experience in an area or from field-wave tests, seismic modeling provides a very affordable approach — especially if we know something about the subsurface geology and the target.

Figure 1 shows ray-tracing examples from Whitmore and Lines (1986) in which raypaths reflected from the flanks of a salt intrusion were used to design the position of the surface source and borehole receivers.

Figure 2 shows wave arrivals that propagate through the model as a function of time. The wavefield snapshots illustrate the complexity of transmitted and reflected wavefields. This modeling also illustrates that the salt-dome reflections also have a dominant horizontal component. This result led to the use of multicomponent recording in the vertical-seismic-profiling experiment for detection of the salt-dome flank.

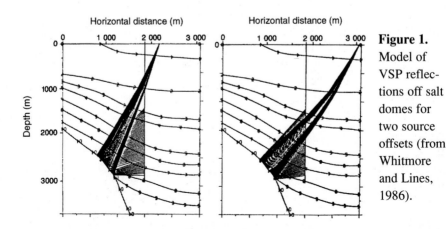

**Figure 1.** Model of VSP reflections off salt domes for two source offsets (from Whitmore and Lines, 1986).

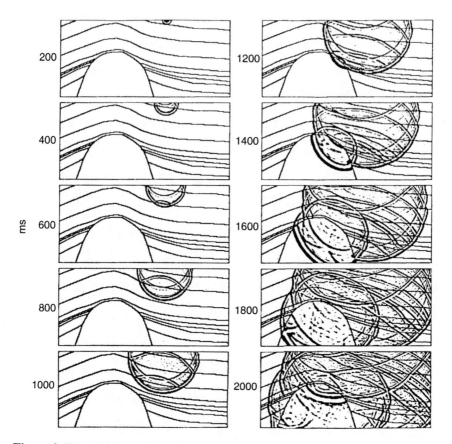

**Figure 2.** Wavefield snapshots for a source at zero offset (from Whitmore and Lines, 1986). Time increments for snapshots = 200 ms.

## Prediction of results

It often has been said in seismic interpretation that if we had not believed it, we wouldn't have seen it. From a modeled seismic data set, we may anticipate the nature of recorded seismic arrivals. Such modeling responses also may provide synthetic data sets with which to test appropriate processing algorithms in an area.

## Enhanced interpretation

To identify reflection events on a seismic section, we generally correlate these events with a synthetic seismogram obtained from the logs of a nearby well. This gives us a starting point for interpretation, and it allows us to determine the polarity of the seismic section. Although the normal-incidence synthetic-seismogram method has been criticized as being simplistic, it remains the main method of tying well data to seismic data. This process requires the science (or art) of choosing a band-limited wavelet that contains the appropriate frequencies for matching the real data. Ideally, both the model and real data contain zero-phase wavelets so that correlating well-log data with real seismic data will represent a process of correlating band-passed reflectivities.

Figure 3 shows an example of correlating a synthetic seismogram with real seismic data. The actual seismic data are on the left side of the figure. The model response (synthetic seismogram) on the right side is obtained by computing reflection coefficients from a sonic log. (Densities in the reflection-coefficient calculations were derived using Gardner's law on sonic velocities.) Model traces are duplicated to give a seismic-section appearance. We note that there is good agreement between the main reflectors of the model and real data sets. On the other hand, noise in the real data clearly distinguishes it from the pristine (noiseless) model data. Despite the noise, a visual correlation will allow us to identify the main reflections in the seismic data.

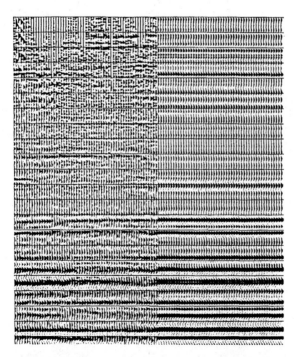

**Figure 3.** Comparison of real data (left) to synthetic seismograms (right) (from Sheriff and Geldart, 1995).

## Inversion

As we will see in later chapters, the procedure of inversion is opposite to forward modeling. Instead of using earth-model parameters and computing a synthetic seismogram, as in the case of modeling, inversion attempts to use the available seismic data to estimate earth-model parameters. To do an inversion, we invoke a modeling method that is appropriate to the geology.

All processing methods inherently assume some type of model, and it is important to know the generality of these models. For example, seismic deconvolution, along with many statics- and velocity-estimation programs, will assume a layered model that is 1D. Seismic tomography is a more general velocity-analysis method that uses ray tracing through 2D or 3D models.

One of the most interesting imaging methods, which can be considered a type of structural inversion, is seismic migration. Each migration method uses a modeling method to place seismic reflectors in their proper subsurface location. Kirchhoff depth migration uses kinematic ray tracing to position subsurface scattering points. Reverse-time migration uses finite-difference wave-equation methods to position reflectors. Although the former is less expensive, it is somewhat less general than the latter. Comparisons of these methods have been described by Zhu and Lines (1998). In later chapters, we will discuss the details of seismic inversion and migration.

## Testing processing algorithms

To test the validity of a seismic-processing algorithm, we compare the estimated answer to the correct answer. In the case of real-data seismograms, the truth rarely is known until a well is drilled. However, with synthetic data, we know the model, and we can successfully compare the effects of processing with the model parameters. In the case of the salt-dome example from Whitmore and Lines (1986), we examined the ability of depth-migration methods to correctly position the salt-dome flank. The result, as shown in Figure 4, is very encouraging.

## Examining effects of noise

We can model noise as well as signal in the synthesis of seismograms. When testing the robustness of a processing or inversion technique, it is worthwhile to examine the way in which noisy data will affect the performance of the processing algorithm.

DEPTH MIGRATION OF VSP (A)    DEPTH MIGRATION OF VSP (B)    SUM OF DEPTH MIGRATIONS

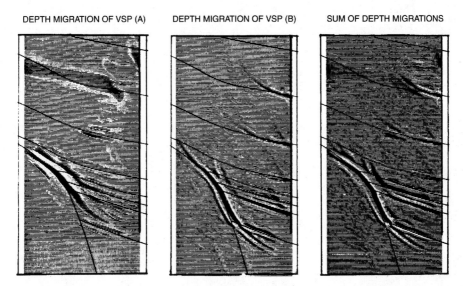

**Figure 4.** Depth migrations of VSP synthetics and their sum (from Whitmore and Lines, 1986).

# Examples of seismic modeling

Seismic acquisition is quite expensive, and the designs of seismic experiments in new exploration areas are economically crucial. One of the primary goals of seismic prospecting is to illuminate interesting areas of the subsurface. We have found ray tracing one of the best ways to do this. If we have some idea of the seismic-velocity model for structures of interest, we can use ray tracing to model the seismic wave paths. This should help us decide on the location and required number of seismic sources and receivers needed to provide adequate subsurface illumination.

Figure 1 shows ray-tracing models for a vertical seismic-profiling experiment near an onshore Gulf Coast salt dome. These particular models helped us in at least three ways. First, they showed the zone of seismic illumination for waves reflected off the salt dome. Second, they gave us some idea of where surface sources should be located for proper illumination. Third, they showed that much of the reflected energy from the dome was traveling horizontally and that optimum recording required multicomponent geophone recording. (The vertical-component phones recorded reflections from sediments, and the horizontal-component phones recorded reflections from the salt flank.)

Although ray tracing is useful for detecting isolated reflections, the complexity of wave propagation is appreciated by examining complete solutions to the wave equation. Figure 2 shows snapshots of wave propagation through the subsurface. A complete solution to the wave equation naturally produces all arrivals, including direct arrivals, refractions, reflections, and diffractions, with all primaries and multiples.

This is the advantage of wave-equation modeling — it can represent every arrival. It is also a disadvantage, because often there are so many arrivals, it is difficult to separate the individual events. In such cases, it is useful to combine full wave-equation modeling with ray tracing.

# References

Aki, K., and P. G. Richards, 2002, Quantitative seismology: University Science Books.

Kelly, K. R., and K. J. Marfurt, 1990, Numerical modeling of seismic wave propagation: SEG.

Sheriff, R., 1991, Encyclopedic dictionary of exploration geophysics: SEG.

Sheriff, R. E., and L. Geldart, 1995, Exploration seismology: Cambridge University Press.

Whitmore, N. D., and L. R. Lines, 1986, Vertical seismic profiling depth migration of a salt dome flank: Geophysics, **51**, 1087–1109.

Zhu, J., and L. R. Lines, 1998, Comparison of Kirchhoff and reverse-time migration methods with applications to prestack depth migration of complex structures: Geophysics, **63**, 1166–1176.

# Chapter 16

# Seismic Inversion

## Overview

*Inversion* can be defined as a procedure for obtaining models that adequately describe a data set. In the case of geophysical data, our observations show the effects of rock properties on physical phenomena such as seismic-wave propagation. Because geophysical inversion allows us to extract geologic-model information from these data, the procedure is of interest to earth scientists.

## Relation between modeling and inversion

The inversion process is related closely to forward modeling, and a comparison between the two procedures is shown in Figure 1. Forward modeling uses a mathematical relationship, such as the wave equation, to synthesize the earth's response for a given set of model parameters. These parameters generally include rock properties and the geometry of rock-layer interfaces.

In seismic modeling, the elastic-wave equation uses the parameters of rock density and wave speed to produce a synthetic seismogram as the model response. In geophysical analysis, it is crucial to choose a forward-modeling procedure that will describe the observations adequately. In addition to the choice of an appropriate mathematical model, it is also important to know how many model parameters should be used and which parameters are significant.

The appropriateness of modeling choices will depend on the particular exploration problem and the geologic area of interest. For example, a model

**Figure 1.** Illustration of the objectives of forward modeling and inversion.

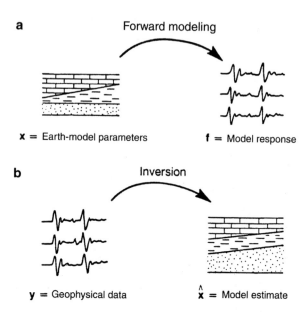

of horizontal layers may be appropriate for geology in central Kansas but would not be suitable in the Overthrust Belt of Wyoming.

As shown in Figure 1b, inversion, or "inverse modeling," uses a "reverse" procedure to that of forward modeling. For a given data set, inversion seeks to define a geologic model that agrees with the observations. Inherent in the inversion process is an attempt to determine the rock-property parameters that allow model responses to fit the available data. Hence, the choice of an appropriate model is important for inversion, and the geophysicist always should be concerned with the physical basis of the inversion model.

Even assuming that we have made the correct modeling choice, numerous problems remain. In fact, Jackson (1972) aptly described inversion in the title of his paper as "interpretation of inaccurate, insufficient, and inconsistent data."

In addressing some of these issues, we adopt a symbolic notation. Thus, we denote the forward-modeling process as a transformation, $\mathbf{f} = T(\mathbf{x})$, where $\mathbf{f}$ is the model response, $\mathbf{x}$ is a vector containing geologic-model parameters, and $T$ is a transformation that mathematically describes a physical process. In the seismic case, $T$ generally produces a model response derived from a solution to the wave equation or an approximation thereof. Then, the process of inversion is written as $\hat{\mathbf{x}} = T^{-1}(\mathbf{y})$, where $\hat{\mathbf{x}}$ is the set of estimated model parameters (the model space) derived from the data vector $\mathbf{y}$ (the data space). The operator $T^{-1}$ denotes the inverse transformation from data space to model space.

Even if the choice of a model (or the choice of T) is physically realistic, certain problems remain with the inverse $T^{-1}$. First, there is a possibility that $T^{-1}$ may not exist. The acquired data may have "blind spots" in which parts of the true subsurface model have not been illuminated by the recordings. Therefore, true subsurface reconstruction may be impossible. Moreover, some inverse problems have more equations than unknowns. Sometimes, such an overdetermined system of equations will be inconsistent and consequently lack a well-posed solution.

On the other hand, many realizations of $\hat{\mathbf{x}} = T^{-1}(\mathbf{y})$ are possible because of nonuniqueness (which also provides employment for those in model fitting). This generally arises as a result of the finite nature of data. Such ill-posed cases could be caused by the presence of more unknowns than equations (an undetermined system), producing many solutions (possibly an infinite number).

This nonuniqueness, or ambiguity, is illustrated by the convolutional model for the seismic trace, as shown in Figure 2a. This trace could be formed by the wavelet and impulse response in Figure 2b, which correspond to a wavelet of positive polarity and a water-layer impulse response for reflections from an ocean bottom with a positive-reflection coefficient. This reflection impulse will exhibit a series of coefficients of alternating signs and diminishing amplitude. (In this example, the reflection coefficient of the water bottom was 0.5).

However, another possible model for the trace in Figure 2a is given by the wavelet and impulse response in Figure 2c. The wavelet is a "ghost" version of the previous wavelet, and the impulse response contains only positive coefficients.

The mathematical similarity of the models in Figures 2b and 2c can be illustrated by use of the $z$-transform. (The $z$-transform associates the coefficients of a time series $-(x_0, x_1, x_2, \ldots, x_n)$ with a polynomial in powers of $z$. In this case, the polynomial would be $x_0 + x_1 z + x_2 z^2 + \ldots + x_n z^n$. This polynomial is equivalent to the discrete Fourier transform for $z = e^{-i\omega}$, for the case in which the temporal sample is unity).

The $z$-transform for the water-bottom reverberation series is given by

$$R(z) = c_1 z^n (1 - c_1 z^n + c_1^2 z^{2n} - c_1^3 z^{3n} + \ldots), \tag{1}$$

**Figure 2.** (a) Trace illustrating the basic mathematical ambiguity of the convolutional model. (b) Model 1: source wavelet (left) and impulse response (right). (c) Model 2: ghosted wavelet (left) and impulse response (right).

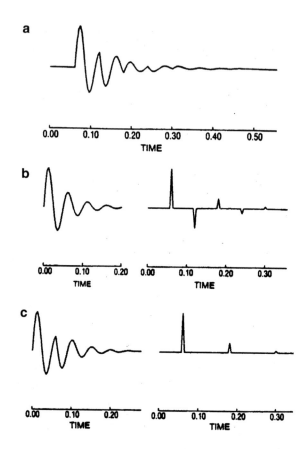

where $n$ is the traveltime index for a water-bottom reflection. If $W(z)$ represents the $z$-transform of the damped sinusoidal wavelet in Figure 2a, the trace can be represented by $Y(z) = W(z)R(z)$. Now $R(z)$ can be factored and written as

$$R(z) = c_1 z^n (1 - c_1 z^n)(1 + c_1^2 z^{2n} + ...).$$  (2)

We see by inspection that if the ghost operator $(1 - c_1 z^n)$ is included in the wavelet instead of the impulse response, we can have

$$Y(z) = W'(z)R'(z),$$

where $W'(z) = W(z)(1 - c_1 z^n)$ and

$$R'(z) = c_1 z^n (1 + c_1^2 z^{2n} + c_1^4 z^{4n}).$$  (3)

$W(z)$ and $R(z)$, respectively, represent the $z$-transforms of the wavelet and impulse responses in Figure 2c. Thus, we have a mathematical dilemma.

Fortunately, we can discard the second model because of the polarity of the arrivals. Because an upgoing pressure wave encounters a negative reflection coefficient of −1.0 at the free surface and the downgoing waves encounter a positive reflection coefficient at the ocean bottom, the multiple reflections have alternating signs. The lesson is that physics often allows us to choose the most credible model from two that initially appear to be valid because they are equivalent mathematically.

The seismic-deconvolution problem is a relevant example for 1D inversion. The uniqueness problems in two and three dimensions also arise but are less severe. (This statement is based not on rigorous proof but on personal experience). Nevertheless, we shall see that such 2D and 3D nonuniqueness problems certainly can arise in kinematic traveltime–inversion problems or in dynamic problems, such as migrations.

A third problem is created because geophysical recordings inevitably are corrupted by noise. Therefore, instead of computing $\hat{x} = T^{-1}(y)$, we compute $\hat{x} = T^{-1}(y + n)$, where $n$ represents the noise or error vector in the measurements. Unfortunately, noise can cause wide variations or instabilities in estimates of the model parameters and can destroy solution validity.

Despite these difficulties, inversion has been used successfully to extract information from geophysical data, as evidenced by examples in the following chapters. The inversion procedure involves various aspects of geophysical analysis, including modeling, processing, and interpretation. Processing steps ultimately attempt to recover a model from the data and therefore may be considered as inversion processes. Because the solutions for limited data sets suffer from nonuniqueness, interpretation is often necessary to choose the most credible model.

The rest of this chapter discusses theoretical aspects and practical applications of the inverse theory in great detail. It will be apparent that despite the limitations of the procedures, inversion has provided geophysicists with valuable information about the earth.

## Seismic inversion using 1D models

Much of seismic-data processing is based on the assumption that local geology can be approximated by a series of horizontal parallel layers. Each of the flat layers in this system possesses characteristic densities, acoustic velocities, and thickness. It was this simple earth model that allowed Dix (1955) to estimate layer velocities from a knowledge of seismic-reflection times and source-receiver distances. The current widespread application of Dix's equation indicates that the 1D model can be appropriate for some data cases. The common-midpoint (CMP) stacking procedure also is based on the layer-cake assumption. In this procedure, it is assumed that summation or stacking of seismic traces that have a common source-receiver midpoint will produce a signal for common-depth-point (CDP) subsurface reflection. An illustration of the model in this assumption is shown in Figure 3.

For such geologic situations, the application of a normal moveout correction, followed by stacking, approximates the response of a normally incident plane wave in a layered medium. The appropriate seismic-trace model for such data is defined by the convolution of a source wavelet with the reflectivity sequence of the medium. The objective of 1D seismic inversion for such data is to estimate the reflection coefficients for these layers in the model. An additional goal often involves estimation of the thickness and acoustic impedance of the layers in the system.

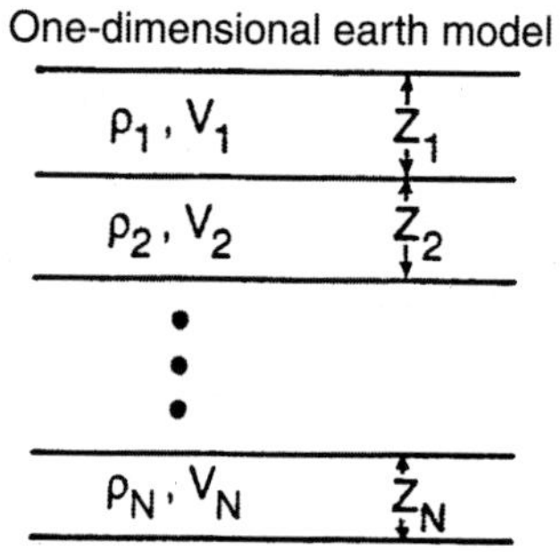

**Figure 3.** One-dimensional earth model consisting of a series of horizontal layers with characteristic densities, velocities, and thicknesses.

One-dimensional earth model

$$\rho_1, V_1 \qquad Z_1$$
$$\rho_2, V_2 \qquad Z_2$$
$$\bullet$$
$$\bullet$$
$$\bullet$$
$$\rho_N, V_N \qquad Z_N$$

The estimation of reflection coefficients for a layered system can be derived by using the Goupillaud earth model, as described in Goupillaud (1961). This model consists of a stratified system in which all layers have equal two-way traveltime intervals. Much of 1D inverse theory is based on this model. Goupillaud's paper demonstrates that the reflection response for a layered system can be considered as a function of the reflection and transmission responses.

Kunetz (1964) used the Goupillaud model to formulate an inversion procedure that derives reflection coefficients for layer interfaces from knowledge of the impulse response for the layered system. Unfortunately, such estimation methods are unstable for noisy data. In practice, reflection coefficients are estimated by a series of processing steps. For real data, reflection coefficients are estimated by attempting to deconvolve source wavelets and multiples from the data. If we assume that deconvolution has been conducted properly, the reflection coefficients for a layered system can be estimated as a function of the two-way vertical traveltime.

# Seismic impedance estimation

Ideally, many geologic interpreters would want to convert estimates of reflection coefficients into an image of acoustic impedance with depth. This inversion problem was addressed in two well-known papers, by Lavergne and Willm (1977) and by Lindseth (1979). Conceptually, the problem would seem to be simple, because for normally incident pressure waves, the reflection coefficients for layer interfaces are related to acoustic impedance by

$$c_k = \frac{\rho_{k+1} \, V_{k+1} - \rho_k \, V_k}{\rho_{k+1} \, V_{k+1} + \rho_k \, V_k},$$ (4)

where $c_k$ = reflection coefficient of the $k$th interface and $\rho_k V_k$ is the acoustic impedance of the $k$th layer. The acoustic impedance can be recovered by the following expression:

$$\rho_{k+1} \, V_{k+1} = \frac{\rho_k \, V_k \, (1 + c_k)}{(1 - c_k)}.$$ (5)

That is, the impedance of a layer can be deduced from the reflection coefficients and the impedance of the layer directly above it. This inversion, often called Seislog® inversion, is actually a nontrivial procedure because seismic data are band limited and noisy. Despite these problems, Lindseth (1979) demonstrates that the Seislog® method provides valuable information that can be used by seismic interpreters.

If reliable density estimates of rocks are known or density-velocity relationships can be assumed safely, the use of seismic-reflection information to derive the impedance variation versus depth can be employed in velocity estimation. Hence, the terms *pseudo-velocity log* (Lavergne and Willm, 1977) and *synthetic sonic logs* (Seislog®) (Lindseth, 1979) describe the attempts to estimate velocity from reflection-coefficient amplitudes.

Before examining the trace-inversion process, we should recognize the motivation for the process. From a seismic point of view, no new information is generated. However, by presenting the seismic data in a different context, we may obtain new interpretational insight. This is especially appealing for geologic interpreters who prefer a sonic log (or pseudosonic log format). Although conventional seismic-reflectivity sections attempt to display the geometry of layer boundaries, the Seislog® section attempts to display parameters, such as impedance or velocity, of the layers between boundaries.

To obtain impedance estimates, we can apply equation 5 recursively. Let $I_k = \rho_k V_k$ denote impedance. If we have an estimate of the acoustic impedance of the top layer, we can find the impedance of deeper layers by repeated application of equation 5. That is, because

$$I_{k+1} = \frac{1+c_k}{1-c_k} I_k , \qquad (6)$$

repeated application gives

$$I_N = I_0 \prod_{i=1}^{N} \frac{1+c_i}{1-c_i} . \qquad (7)$$

For most geologic boundaries, typical reflection coefficients rarely exceed 0.3. That is, $c_k \leq 0.3$ and, to a good approximation, equation 4 can be expressed as

$$c_k = \frac{\nabla I_k}{2 I_k} . \qquad (8)$$

If we regard the reflectivity sequences as a continuous sequence, we can say that in lim $\Delta t \to 0$,

$$c(t)dt = \frac{1}{2} d(\ln(I(t))) . \qquad (9)$$

Integration over time yields

$$I(t) = I_0 \exp \left[ 2 \int_0^t c(\tau)d\tau \right] . \qquad (10)$$

In many cases, we may wish to deal with seismic velocities rather than impedance, because fractional velocity changes are usually greater than fractional density changes. Moreover, density variations for certain rock types may obey Gardner's relations between density and velocity (Gardner et al., 1974). That is, density ($\rho$) may be expressed in terms of velocity, $v$, as

$$\rho = \beta v^{\alpha} . \qquad (11)$$

For Gardner's relationship, $\alpha = 0.25$ and $\beta = 0.23$ (for velocity in ft/s and density in gm/cm$^3$). In some cases, density is assumed to be constant, in which case $\beta = 1$ and $\alpha = 0$.

Upon substitution for density and velocity from equation 11, we obtain

$$V(t) = V_0 \exp\left[\frac{2}{\alpha+1}\int_0^t c(\tau)d\tau\right]. \tag{12}$$

Hence, the velocity function can be obtained by integration of the reflectivity series over time.

Although the mathematics of this reflectivity-integration method are fairly straightforward, several assumptions have been made implicitly. It has been assumed that we can obtain the reflectivity sequence from the data by processing. That is, we assume that the data have been deconvolved properly to remove the effect of multiples and the source wavelet. An ideal wavelet deconvolution would remove distortions of the wavelet's phase and amplitude spectrum. The latter is equivalent to the assumption that we have a full-band zero-phase wavelet. In addition, this inversion process assumes a high signal-to-noise ratio in our seismic traces.

For interpretation of seismic inversions in terms of geologic depths, it is also necessary to have the data migrated to the correct lateral positions.

In addition, there is the issue of trace scaling in seismic inversion. As is evident from equation 4, reflection coefficients vary in amplitude from $-1$ to $+1$. On the other hand, seismic-trace amplitudes are meaningful only in their relative sizes, and numerical amplitudes undergo adjustment in processing. The nontrivial issue is to process seismic data so that amplitudes are scaled to a range in which they represent the amplitudes of seismic-reflection coefficients.

The most crucial problems of the Seislog® impedance estimation arise because of scaling problems and the band-limited nature of the seismic data. These difficulties were recognized by Lindseth (1979), who showed examples that outlined the problems of missing frequencies and scaling.

A major difference between sonic logs and synthetic sonic logs produced from seismic data by Seislog® methods is that low-frequency bands (typically 0–5 Hz) are missing in the seismic data. An example from Lindseth (1979) illustrates this feature.

Figure 4 shows the effect of removing the low-frequency band 0–5 Hz from a time-sampled sonic log in which the Nyquist frequency is 250 Hz. The sonic log is shown to be the sum of the detailed velocity function

(6–250 Hz) and the gross velocity function (0–5 Hz). This missing low-frequency band is typical of that missing from seismic data. Therefore, the figure essentially shows that even an idealized inversion (with the entire high-frequency band) would be missing the important trend information contained in sonic logs.

Another major problem is the missing high-frequency component of the impedance, because high frequencies (typically 100 Hz to the Nyquist frequency) also are missing from seismic data. Figure 5 (from Lindseth, 1979)

**Figure 4.** Example showing that a sonic log may be expressed as the sum of a gross velocity function (0–5Hz) and a detailed velocity function (6–250 Hz) (from Lindseth, 1979).

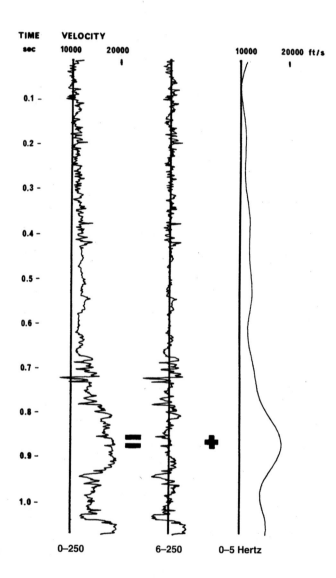

illustrates this problem by showing low-pass-filtered sonic logs in which each low-pass filter has a different high-frequency cutoff. As the cutoff is lowered, resolution is decreased. The worst-case situation of Figure 5 has a cutoff frequency of 50 Hz, and this is not unusual for many seismic data sets.

A mathematical analysis of the missing low-frequency band can be done by considering the following decomposition of reflectivity into low-frequency and high-frequency components:

$$c(t) = c_L(t) + c_H(t), \tag{13}$$

where $c_L(t)$ is missing the low-frequency reflectivity, $(0 - fL$ Hz), and $c_H(t)$ is seismic reflectivity ($fL - fH$ Hz). Here, $fL$ is the low-frequency cutoff for the seismic data and $fH$ is the high-frequency cutoff.

In this case, substitution into equation 10 yields

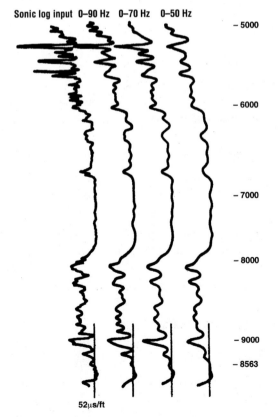

**Figure 5.** Example showing that the removal of high-frequency components from sonic logs results in decreased resolution (from Lindseth, 1979).

$$I = I_0 \exp2\int(c_L + c_H)dt =$$

$$I_0 \underbrace{\exp(2\int c_L dt)}_{IL} \underbrace{\exp(2\int c_H dt)}_{IH}. \tag{14}$$

From equations 13 and 14, we see that there are essentially two methods of inserting low-frequency information into equation 14 once $c_L(t)$ is known. We can apply either equation 13 or 14. That is, we can compute the reflectivity in low-frequency and high-frequency parts, as shown in equation 13, before applying the impedance calculations of equation 5. Alternatively, we can apply equation 14 to compute impedance by multiplying $I_H$ by $I_L = Io$ $\exp 2\int c_L$ dt.

Such approaches require that we find some method of retrieving low-frequency functions $c_L(t)$ or $I_L(t)$ This normally is achieved by using existing sonic-log information or by using estimates of interval velocity (perhaps obtained from stacking velocity).

## Interpretation examples

Many of the first applications of impedance estimation were shown by Lindseth (1979). These applications generally were related to detection of stratigraphic traps, including zones of carbonate and sandstone porosity. Zones of increased porosity usually are related to zones of lowered acoustic impedance.

One of these applications is shown in Figures 6a and b. The seismic section for the line in Figure 6a indicates a slight decrease in amplitude in the Crossfield carbonate (part of the Devonian Wabamun Formation) between shotpoints (SPs) 39 and 55. Lindseth's Seislog® results in Figure 6b show this lowering of acoustic impedance in the Crossfield between SPs 39 and 55, compared to SPs 15–39. These zones of lowered impedance corresponded to porous gas-filled carbonate.

The porosity was indicated more clearly on the Seislog® sections than on the original seismic sections.

The method of seismic-impedance estimation also can be applied in the evaluation of lithology by inverting both P-wave and S-wave data. This was shown in an award-winning talk presented in 1990 by Gary Ruckgaber at the Norwegian Petroleum Institute Meeting. Ruckgaber used multicompo-

**Figure 6.** (a) Real-data example showing a seismic section in which there is production from the Crossfield carbonate (from Lindseth, 1979). (b) Synthetic sonic logs for data in Figure 4b. The Seislog inversion of Lindseth (1979) showed an increase in slowness (and porosity) for the Crossfield carbonate between shot-points 39 and 55.

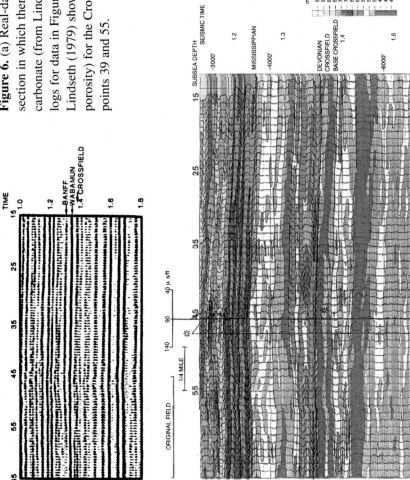

nent data from Hambert Field, Colorado, recorded with P-wave and shear-wave vibrators. The seismic trace inversion worked in the same manner for both P-wave and S-wave data (Ruckgaber, 1990).

The advantage of multicomponent recording over vertical component recording is that multicomponent data allow us to estimate $V_P/V_S$ ratios in addition to P-wave velocity. This has the advantage of allowing us to distinguish between lithologies such as sandstone and shale, as we will investigate further in Chapter 22.

# References

Dix, C. H., 1955, Seismic velocities from surface measurements: Geophysics, **20**, 68–86.

Gardner, G. H. F., L. W. Gardner, and A. R. Gregory, 1974, Formation velocity and density — The diagnostic basics for stratigraphic traps: Geophysics, **39**, 770–780.

Goupillaud, P. L., 1961, An approach to inverse filtering of near surface layer effects from seismic records: Geophysics, **26**, 654–670.

Jackson, D. D., 1972, Interpretation of inaccurate, insufficient, and inconsistent data: Geophysical Journal of the Royal Astronomical Society, **28**, 97–109.

Kunetz, G., 1964, Generalization des operateurs d'antiresonance a un nombre quelconque de reflecteurs: Geophysical Prospecting, **12**, 283–289.

Lavergne, M., and C. Willm, 1977, Inversion of seismograms and pseudo velocity logs: Geophysical Prospecting, **25**, 231–250.

Lindseth, R. O., 1979, Synthetic sonic logs — A process for stratigraphic interpretation: Geophysics, **44**, 3–26.

Ruckgaber, G., 1990, Seismic lithology estimation at Hambert field, Colorado, U.S.A.: Presented at the 1990 Norwegian Petroleum Institute Meeting.

Treitel, S., and L. R. Lines, 1988, Geophysical examples of inversion (with a grain of salt): The Leading Edge, **7**, 32–35.

# Chapter 17

# Seismic Traveltime Tomography

The word *tomography* comes from the Greek *tomos*, meaning to cut or slice. Thus, tomography is based on the premise that observed data sets are related to line integrals along lines or rays (i.e., projections) of some physical quantity. Tomography is used to reconstruct a model of the desired physical object so that the model's projected data agree approximately with measurements. A classic geophysical tomography problem is the reconstruction of a seismic velocity model of some portion of the earth in which the computed traveltimes agree with the observed traveltimes. In other words, traveltime tomography is a procedure that allows us to invert observed seismic traveltimes to estimate the subsurface velocity structure.

In geophysical tomography, the focus has been on the seismic-traveltime inversion problem. In seismic-traveltime tomography, we must consider two distinct cases. First is the reflection problem in which both source and receiver are at the surface of the earth. Second is the transmission problem in which the sources and/or receivers are in boreholes beneath the surface. Hybrid problems, such as VSP (vertical seismic profiling), in which both reflection and transmission are important, represent a straightforward generalization of these two cases. In any event, the essence of traveltime tomography is that the traveltime associated with a given ray is the integrated slowness along the ray. In the 2D case, one has

$$t(\text{ray}) = \int_{raypath} s(x,z)dl , \qquad (1)$$

where $x$ and $z$ are horizontal and vertical coordinates, $dl$ is the differential distance along the ray, and $s(x, z) = 1/v(x, z)$ is the slowness (reciprocal velocity) at the point $(x, z)$.

A fundamental difficulty of traveltime tomography is that the raypath itself depends on the unknown slowness distribution; therefore, equation 1 is nonlinear. The approach traditionally used in tomography is to linearize equation 1 about some initial, or reference, slowness model. When this is done, the discrete form of the linearized integral equation, for a collection of rays, is

$$\Delta t = D \Delta s, \tag{2}$$

where $\Delta t$ is a vector whose components are the difference between the traveltimes computed for the model and the observed traveltimes, $\Delta s$ is a vector whose components are the differences in slowness between the model and the true solution, and $D$ is a matrix whose element $D_{ij}$ is the distance the $i$th ray travels in the $j$th cell (Figure 1).

Various discretizations of the model are possible. To simplify the discussion, we can assume that the model is covered with square cells of constant size, within which the slownesses are constant. More sophisticated methods (e.g., Langan et al., 1985) allow for velocity gradients within cells.

The first step in the tomography procedure is to obtain traveltimes from unstacked seismic sections. This can be done by traveltime picking on an interpretive workstation or from direct field measurements (as in Gustavsson et al., 1985). Rays are then traced through the original model. The rays correspond to all the events for which traveltimes have been picked. The simplest ray tracing ignores Snell's law at all cell boundaries, so that sources and receivers are joined by straight lines and the distances $D_{ij}$ are computed by simple trigonometry.

A more realistic ray-tracing procedure bends the rays at each cell boundary according to Snell's law, and the distances $D_{ij}$ are again computed easily. In the case of reflection tomography, allowance must be made for a ray to be reflected at geologic interfaces. Strictly speaking, rays could reflect from every cell boundary, but it will be assumed that velocity contrasts at the cell boundaries are small enough that such reflections can be neglected. Reflections will be allowed to occur only at prescribed interfaces (Cassell, 1982; Langan et al., 1985).

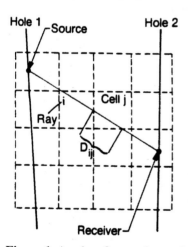

**Figure 1.** A subsurface region and its discretization into cells (after Dines and Lytle, 1979). The $i$th ray travels a distance $D_{ij}$ in the $j$th cell.

The *i*th row of the **D** matrix describes the distance of the *i*th ray in the cells that the ray traverses. A value of 0 for the *j*th column indicates that the ray does not penetrate the *j*th cell. The number of rows equals the number of rays, and the number of columns is equal to the number of cells used to describe the model. In general, **D** will have more rows than columns, because one wants the model to be illuminated with as many rays as possible. This means that equation 2 usually will be overdetermined because there are usually more rays (rows) than cells (columns). Accordingly, a least-squares solution to equation 2 is computed.

The solution is a set of slowness perturbations, $\Delta s$, that are added to the initial slowness model to produce an updated slowness model. At this stage, the procedure can be repeated by using the updated slowness field as a new starting model. Producing an updated model requires tracing rays through the updated model in which Snell's law is to be satisfied. In the case of straight rays, the distance elements, $D_{ij}$, remain fixed and need to be determined only once.

In either case, traveltimes must be computed for the updated slowness model. This iterative process can continue until some previously established stopping criterion has been satisfied. The set of cell velocities obtained after each iteration is called a tomogram.

One formulation of reflection tomography distinguishes between the nonreflecting cell boundaries and the explicitly defined subsurface reflecting interfaces (Figure 2). These reflectors can be expressed either analytically or numerically, e.g., as splines. The starting model consists of the initial cell slowness along with a set of initial reflecting interfaces. Bishop et al. (1985) perform least-squares inversion for both cell velocities and inter-

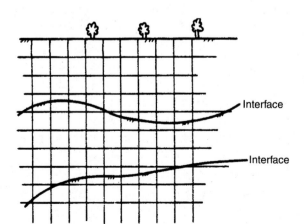

**Figure 2.** Parameterization of the subsurface in terms of velocity cells and interfaces (from Bording et al., 1987). In the inversion, the velocities are assumed to be constant in cells.

face geometry. Williamson (1984), however, assumes that the reflecting boundary positions are known and uses tomography only to infer the velocity field.

An alternative approach, proposed by Stork and Clayton (1985) and by Whitmore and Lines (1986), uses an updated velocity model from tomography as the input for a depth-migration scheme. The repeated use of traveltime tomography followed by depth migration is called iterative tomographic migration. This dual procedure is repeated until a previously established stopping criterion has been satisfied.

We favor iterative tomographic migration because a good feature of tomography — velocity inversion — is combined with a good feature of depth migration — imaging of the interfaces. In fact, the good performance of a depth-migration scheme is predicated on good knowledge of the subsurface velocity distribution, and thus the procedures complement each other nicely. Before showing examples of traveltime tomography, the component procedures will be described in some detail.

## Traveltime interpretation

Traveltimes are the data for traveltime tomography, and unless these are measured in the field, they must be picked from the recorded seismograms prior to the stacking process. The advent of modern interpretive workstations has made this task tractable. Prestack-traveltime interpretation involves digitizing across horizons on the time sections. These horizons are the trace-to-trace coherent patterns associated with reflections from the subsurface boundaries.

Although common source records were used originally (Bording et al., 1987), there are advantages to picking in the common-offset domain. First, common-offset records have time variations caused by geology, not by normal-moveout (NMO) effects of source-receiver geometry. Hence, the interpreter does not follow NMO hyperbolas on common-offset gathers but simply examines event delays between records.

As a starting point, the times of the near-offset record can be obtained from an initial stacking of the data. Then, the events on records of greater offset can be tracked. Care is needed to minimize traveltime errors, because traveltimes are essentially the data that we invert. Traveltime picking is an important application for seismic workstations, and automatic picking procedures are applied on them routinely.

# Ray tracing

Tomography is based on the assumption that seismic-wave propagation can be described accurately by ray tracing. This allows the wave equation — the partial-differential equation describing energy propagation in the medium — to be reduced to a partial differential equation for the wavefront (Aki and Richards, 2002). Modeling the energy propagation through the medium by ray tracing amounts to solving the ray equation for a given background-velocity model, a set of reflecting boundaries, and a collection of source/receiver pairs. We assume that rays reflect only from the specified reflecting boundaries and not from velocity jumps resulting from discretization of the media bounded by the reflectors.

The ray-tracing codes commonly used in many reflection-tomography applications are based on the algorithm of Langan et al. (1985) and of Stork (1988). In this algorithm, the velocity model is discretized into a mesh of velocity points in which constant velocity gradients are allowed with each cell. Because the velocity gradient in each of the triangles making up the cell is constant, the path of the ray across the triangle is an arc of a circle (see, for example, Slotnick, 1959, p. 205).

# Tomography inversion

After traveltimes have been picked and rays have been traced through a model, one must solve a large, usually overdetermined system of traveltime equations for the unknown slowness perturbation vector (equation 2). Recall that the elements of the distance matrix, **D**, correspond to the distance that a ray travels through a particular cell. Because a given ray intersects only a very small portion of the model, most of the elements in that row are zero. Thus, the matrix **D** is very sparse. In fact, the more cells a model has, the sparser it tends to be.

The reflection-tomography example to be shown is approximately 99% sparse. By taking advantage of this sparsity, one can achieve huge improvements in run time and memory requirements. Therefore, sparse iterative techniques, such as preconditioned conjugate gradient (Scales, 1985) and LSQR (Paige and Saunders, 1982), are used. Row-action methods such as ART (algebraic-reconstruction technique) and SIRT (simulation iterative reconstruction technique) (Herman, 1980), described below, also can be adapted for sparsity, but they tend to converge more slowly (Scales, 1987).

ART and SIRT are popular row-action methods because they deal with individual rays (matrix rows) of large systems of equations, as outlined by Lytle et al. (1978). The basic principle behind these two methods is quite simple.

In ART, we consider the $i$th traveltime equation from system 2 given by

$$\sum_j D_{ij}\Delta s_j = \Delta t_i. \tag{3}$$

The slowness update is assumed to be proportional to the amount of the $i$th raypath in cell $j$ relative to the total raypath. That is,

$$\Delta s_j = \frac{\alpha_i D_{ij}}{\sum_j D_{ij}}, \tag{4}$$

where $\alpha_i$ is a proportionality constant whose value is given by back-substitution into equation 3, so that

$$\alpha_i = \frac{\Delta t_i \sum_j D_{ij}}{\sum_j D_{ij}^2}. \tag{5}$$

This component is made for the full set of raypaths and is repeated until convergence is obtained.

SIRT is a very similar method except that it operates on all paths passing through a given cell. The expression is averaged for all raypaths that have passed through that cell. That is, updates to the slowness values are made after computing all raypaths through a cell. SIRT has the advantage over ART in that the solution is independent of equation ordering.

Having introduced the concepts of reflection and transmission tomography, we now examine specific types of traveltime tomography.

# Surface-to-borehole tomography and VSP imaging

As indicated previously, transmission tomography includes surface-to-borehole as well as borehole-to-borehole transmission (Figure 3). Surface-to-borehole tomography is a special case of vertical seismic profiling (VSP). The application of surface-to-borehole traveltime tomography methods is not new to geophysics. Various papers in earthquake seismology (for example, Aki and Richards, 2002) have dealt with the problem of earthquake or explosive sources at depth with receivers at the surface.

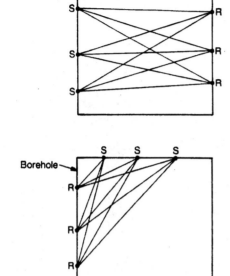

**Figure 3.** Typical borehole-to-borehole and surface-to-borehole geometries (after McMechan, 1983). The top geometry shows cross-borehole recording and the bottom shows a VSP. S = source; R = receiver.

Various parameterizations of the model have been used in VSP traveltime inversion, as shown by McMechan (1983), by Stewart (1984), and by Lines et al. (1984). These studies show that VSP traveltime inversion works well under the following conditions:

- Traveltimes must be accurate. For good data, it generally is conceded that the data sample interval is an honorable goal for picking accuracy as well as a criterion for traveltime data fitting.

- The collection of all raypaths must have sufficient angular coverage (aperture). A wide range of ray-incident angles through the medium is desirable. Hence, wide offset range is recommended for recording.

- It is desirable to invert VSP reflection traveltimes in addition to transmitted arrivals.

After the estimation of velocity by traveltime inversion, the subsurface can be imaged effectively by seismic depth migration. These subsurface velocity models should produce transmission and reflection traveltimes that match all available data. Figure 4 (from Whitmore and Lines, 1986) shows reflected VSP rays for a salt-dome model, which was part of a traveltime-tomography case history.

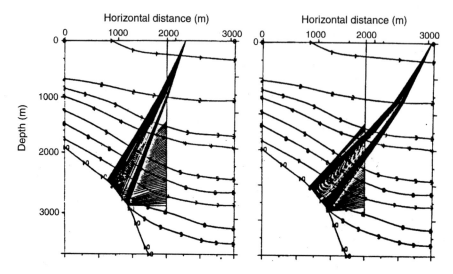

**Figure 4.** Model of VSP reflections off a salt dome for two source offsets (from Whitmore and Lines, 1986).

# Cross-borehole tomography

Coinciding with the ongoing developments of downhole seismic sources and seismometers and because of possible shortcomings of conventional exploration, cross-borehole seismic exploration recently has become a research topic of interest. Conventional reflection seismology records elastic waves that have been initiated by a near-surface source. However, for reflection seismology, significant high-frequency attenuation can occur in near-surface weathering layers. Such absorption problems are lessened in cross-borehole seismic methods by placing sources and receivers below the near-surface layers (Dines and Lytle, 1979; Ivansson, 1986; Macrides et al., 1988). The recording geometry for such surveys is shown in Figure 3.

Velocity control is available from sonic logs at a borehole. An objective of geologic interpretation is to estimate velocities between boreholes by interpolation. However, an approach preferable to interpolation involves a cross-borehole survey in which seismic traveltime measurements allow the interborehole velocity to be estimated by cross-borehole tomography. Examples of this application in reservoir studies are given by Lines (1991).

# Refraction tomography

One of the most relevant data-processing problems for land seismic data involves traveltime corrections for the surface weathering layer. This layer often exhibits a high degree of variability in velocity, causing large time shifts ("statics") in reflections, thereby creating apparent structures in the seismic time section. Effective refraction traveltime-inversion solutions for the problem have been proposed by Hampson and Russell (1984), using a layered velocity model, and by Docherty (1992), using cell-based tomography. Both procedures model the head-wave arrivals, which are refracted critically along the base of weathering.

The Docherty model describes the weathering layer by a series of velocity cells in which the cell depth is described by a set of nodal points. This model allows for lateral velocity variation within the weathering layer and for a varying depth of the layer. The fitting of traveltimes by adjusting velocities in the model allows for the accurate calculation of weathering traveltime corrections. The effectiveness of the solution is demonstrated by Docherty (1992), who compared traveltime tomographic-processed data to conventional processing. As shown in Figure 5, reflector continuity in the tomography processing appears to be better than the stacked section obtained from conventional methods.

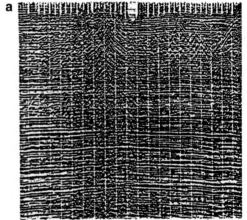

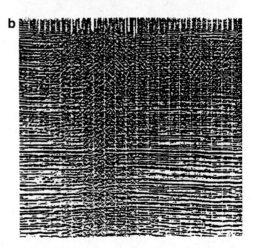

**Figure 5.** Data examples (from Docherty, 1992) using refraction tomography (a) and conventional processing (b).

# Future possibilities

Traveltime tomography produces favorable results for reflection, cross-borehole, and refraction tomography. Tomographic methods often benefit from not using restrictive assumptions.

Although the applications shown here have been generally 2D in nature, 3D reflection traveltime inversion has been applied to enhance oil-recovery problems, as shown by Lines et al. (1990).

Traveltime tomography shows promising possibilities for velocity analysis of surface reflection and borehole seismic recordings. The method can be coupled effectively with depth migration for both surface and borehole seismic data recordings. If seismic traveltimes can be obtained effectively, we anticipate many future applications of seismic tomography.

# References

Aki, K., and P. G. Richards, 2002, Quantitative seismology: University Science Books.

Bishop, T., K. Bube, R. Cutler, R. Langan, P. Love, J. Resnick, R. Shuey, D. Spindler, and H. Wyld, 1985, Tomographic determination of velocity and depth in laterally varying media: Geophysics, **50**, 903–923.

Bording R. P., A. Gersztenkorn, L. R. Lines, J. A. Scales, and S. Treitel, 1987, Applications of seismic traveltime tomography: Geophysical Journal of the Royal Astronomical Society, **90**, 285–303.

Cassell, B., 1982. A method for calculating synthetic seismograms in laterally varying media: Geophysical Journal of the Royal Astronomical Society, **69**, 339–354.

Dines, K. A., and R. J. Lytle, 1979, Computerized geophysical tomography: Proceedings of the Institute of Electronic and Electrical Engineers, **67**, 1065–1073.

Docherty, P., 1992, Solving for the thickness and velocity of the weathering layer using 2-D refraction tomography: Geophysics, **57**, 1307–1318.

Gustavsson, M., S. Ivansson, P. Moren, and J. Pihl, 1985, Seismic borehole tomography measurement system and field studies: Proceedings of the Institute of Electronic and Electrical Engineers, **74**, 339–346.

Hampson, D., and B. Russell, 1984, First-break interpretation using generalized linear inversion: Canadian Society of Exploration Geophysicists Journal, **20**, 40–54.

Herman, G., 1980, Image reconstruction from projections: Academic Press.

Ivansson, S., 1986. Seismic borehole tomography-theory and computational methods: Proceedings of the Institute of Electronic and Electrical Engineers, **74**, 328–338.

Langan, R., I. Lerche, and R. Cutler, 1985. Tracing rays through heterogeneous media: An accurate and efficient procedure: Geophysics, **50**, 1456–1466.

Lines, L. R., 1991, Applications of tomography to borehole and reflection seismology: The Leading Edge, **10**, no. 7, 11–17.

Lines, L. R., A. Bourgeois, and J. D. Covey, 1984, Traveltime inversion of vertical seismic profiles — A feasibility study: Geophysics, **49**, 250–264.

Lines, L. R., R. Jackson, and J. D. Covey, 1990, Seismic velocity models for heat zones in Athabasca tar sands: Geophysics, **55**, 1108–1111.

Lytle, R., K. Dines, E. Laine, and D. Lager, 1978, Electromagnetic cross-borehole survey of a site proposed for a future urban transit station: Lawrence Livermore Laboratory Report UCRL-52484.

Macrides, C. G., E. R. Kanasewich, and S. Bharatha, 1988, Multi-borehole seismic imaging in steam injection heavy oil recovery projects: Geophysics, **53**, 65–75.

McMechan, G., 1983, Seismic tomography in boreholes: Geophysical Journal of the Royal Astronomical Society, **74**, 601–612.

Paige, C., and M. Saunders, 1982, LSQR: An algorithm for sparse linear equations and sparse least squares: Transactions of the Association of Computational Mathematics on Mathematical Software, **8**, 43–71.

Scales, J., 1987, Tomographic inversion via the conjugate gradient method: Geophysics, **52**, 179–185.

Slotnick, M., 1959, Lessons in seismic computing: SEG.

Stork, C., and R. Clayton, 1985, Iterative tomographic and migration reconstruction of seismic images: Paper presented at the 55th Annual International Meeting, SEG.

Stewart, R. R., 1984, VSP interval velocities from traveltime inversion: Geophysical Prospecting, **32**, 608–628.

Stork, C., 1988, Ray trace tomographic velocity analysis of surface seismic reflection data: Ph.D. thesis, California Institute of Technology.

Whitmore, N. D., and L. R. Lines, 1986, VSP depth migration of a salt dome flank: Geophysics, **51**, 1089–1109.

Williamson, P., 1984, The application of tomography to the inversion of traveltime data in reflection seismology: Paper presented at the 46th Annual Meeting, European Association of Geoscientists and Engineers.

# Chapter 18

# 3D Reflection Seismology

We live in a world of three spatial dimensions. This is crucial to remember when we examine geologic plays and petroleum traps. We interpret individual seismic lines as though the earth were varying in two dimensions beneath these profiles, but we do so with the risk that "out-of-plane" reflections can occur. To solve the problem of 3D geologic interpretations, the industry now uses areal arrays of sources and receivers at the surface rather than widely spaced profiles. These seismic surveys allow us to accurately image 3D spatial variations in geology beneath the earth's surface. Brown (1991) and Liner (1999) have given excellent summaries of 3D interpretation and its importance.

## Motivation

One of the first examples that demonstrated the need for 3D imaging was the "French double-dome model," published by French (1974). This physical model, which simulated seismic surveys over two modeled domes near a faulted surface, is shown in Figure 1. This figure shows a series of 13 seismic lines that were shot over the dome. The raw unmigrated data for profile 6 is shown in the figure.

Reflection energy is coming not only from the dome beneath the line but also from reflections off the nearby dome and the fault surface. If we migrate this line as an individual profile, considering that all energy comes from beneath the line, we obtain the 2D migration shown in the middle seismic section. We note that some out-of-plane artifacts are created in the migrated image. A better result is obtained if we consider all the seismic data and migrate the data into a 3D volume of seismic data while using a 3D velocity model. This 3D migration produces a superior result, as shown in Figure 1, albeit at greater expense.

**Figure 1.** "French double-dome" model (modified from French, 1974) accompanied by unmigrated data, 2D migration, and 3D migration.

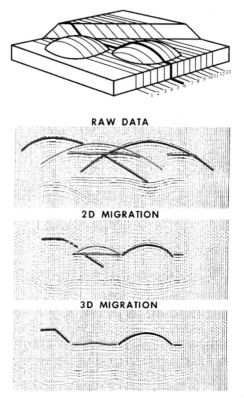

RAW DATA

2D MIGRATION

3D MIGRATION

# 3D seismic surveys

The lesson to be learned from this example is that when we deal with rapid 3D variation of the subsurface, we must focus on 3D data recording and 3D imaging methods. In recent years, several books on 3D survey design have appeared. These include publications by Stone (1994) and by Cordsen et al. (2000). An unpublished set of course notes for the Canadian Society of Exploration Geophysicists, by Norman Cooper (1995), also give lucid explanations of 3D survey design. We follow some of his discussions here.

In many ways, the definitions of 3D fold and common-midpoint bins are similar to their 2D equivalents. Figure 2 shows the definition of a common midpoint (CMP) or common depth point (CDP). (For flat reflectors, CMP = CDP.) The CMP or CDP is defined as the reflecting point on the subsurface reflector whose horizontal distance is halfway between the source and the receiver. Figure 2 is a reminder that the subsurface area is half the size of the surface coverage. Fold is given by the number of traces from various source-receiver combinations that sample the CMP.

For 3D surveys, we use the same definitions of CMP, fold, and coverage. The main difference is that for 3D surveys, we deal with areal arrays of sources and receivers. The CMP is defined by taking the midpoint of the $x$ and $y$ coordinates for the sources and receivers rather than the midpoint in the $x$ direction. Figure 3 shows a diagram of source-and-receiver arrays and how they can combine to give an areal response. As shown in Figure 2, the subsurface coverage is half the surface dimensions in $x$ and $y$, meaning that the areal coverage at depth is one-quarter that of the surface areal arrays.

The fold is computed in the same manner as before, except that we deal with the number of traces that sample subsurface areal bins. Figure 4 shows the subsurface coverage of a single shot fired into an array of surface re-

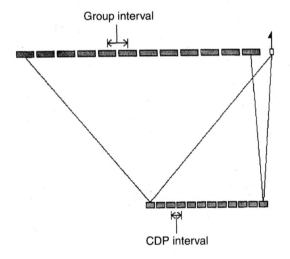

Group interval

CDP interval

**Figure 2.** Definition of a common depth point (CDP) (from Cooper, 1995). It serves as a reminder that the subsurface coverage is half the surface distance.

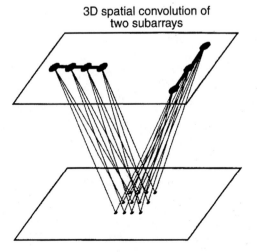

3D spatial convolution of two subarrays

**Figure 3.** Areal arrays of sources and receivers with subsurface coverage (from Cooper, 1995).

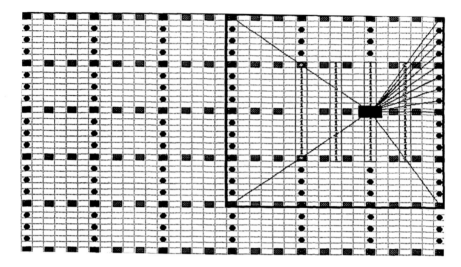

**Figure 4.** Fold coverage for one shot fired into the receiver patch (from Cooper, 1995).

ceivers. Shot locations are marked by squares, and receiver locations are marked by circles. The present shot location is given by the large black rectangle. For this shot, a patch of receivers outlined by the black rectangle is turned on for the shot. The locations of bins that are illuminated by the experiment are denoted by the number 1.

If we repeat the experiment for several shots, we will add to the fold number for every time the bin is a CMP for a particular shot-receiver combination. Our coverage, or fold diagram, will be given by the number of traces for each bin.

# A simple example of why we need 3D

Wu et al. (1996) showed a very simple example of why 3D imaging is necessary. The authors constructed a model of two-point diffractors in different vertical planes. One point had $(x, y, z)$ coordinates of (800, 400, 1600) and the other had coordinates of (1600, 800, 800).

Figure 5 shows two surface seismic profiles for $y = 400$ m and 800 m. We see the diffraction hyperbolas in both these lines. It is not apparent that in each case, one of the hyperbolas comes from out-of-plane energy. If we were to migrate these lines using 2D migration, we would see the incorrect answer of diffraction points in both sections, as shown in Figure 6. The 2D migration cannot recognize out-of-plane energy. However, if we take

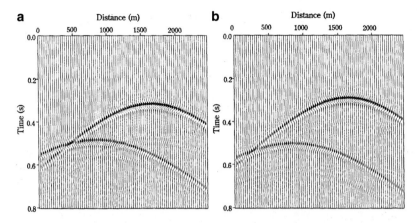

**Figure 5.** Diffraction hyperbolas for two-point diffractors (from Wu et al., 1996); (a) and (b) represent the planes $y = 400$ m and $y = 800$ m, respectively.

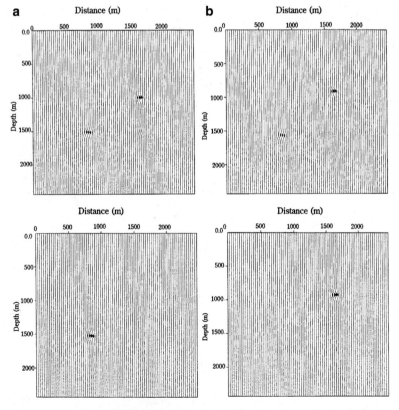

**Figure 6.** Contrast of 2D migrations (top) and 3D migrations (bottom) for the two-point diffractor model (from Wu et al., 1996). Columns (a) and (b) represent the planes $y = 400$ m and $y = 800$ m, respectively.

recordings from the entire area of the top surface and apply 3D migration, we note that the diffraction points are in their correct locations. The correct answer comes from using 3D acquisitions and imaging (migration).

# 3D imaging of real data

Figure 7 shows that 3D migration does a superior job on complex structures, compared to 2D migration. The out-of-plane reflections have caused some blurring in 2D migration, but the 3D migration shows superior focusing.

We see similar comparisons for the imaging of salt intrusions in the Gulf of Mexico, in a case history of Ratcliff et al. (1996). As we have shown in Chapter 14 of this book (on plays related to salt structures), the failure to recognize the 3D nature of the salt play resulted in the drilling of a dry hole.

If we use accurate velocity models, 3D imaging almost always will do a better job than 2D imaging. The only time they would do equally well would be in the case of a structure that could be considered "two-dimensional." However, even if structures are considered to be 2D in the dip direc-

**Figure 7.** Comparison of unmigrated, 2D-migrated, and 3D-migrated sections, demonstrating the need for 3D imaging (from Brown, 1991).

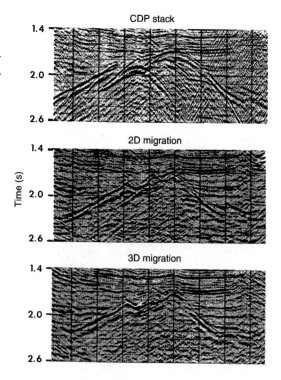

tion with little variation in the strike direction, 3D imaging often does a better job. An example of this is shown by Wu et al. (1996), in which images for an erosional valley structure are compared.

## Time and depth slices

In previous chapters, we have viewed seismic data in cross sections, as though we could take a vertical slice through the earth. With a 3D volume of seismic data, we can use horizontal slices through the earth. These views are termed time slices (or depth slices if data are depth migrated). These images can show geology as it may have existed at an earlier period of geologic history. This is especially useful if we are examining salt intrusions or sand channels. Figure 8 shows such a meandering stream channel from a Thailand survey. The geology is well defined by the time slice itself.

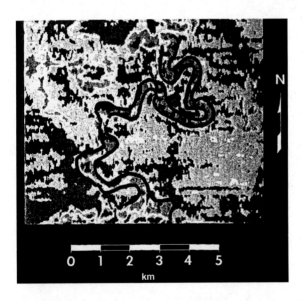

**Figure 8.** 3D image of a meandering stream (from Brown, 1991).

## The effects of 3D methods on interpretation

Given the previous comparisons of 2D and 3D images for models and processed data, it is evident that 3D methods should have a beneficial effect on interpretation. This issue is addressed in a paper by Nestvold (1992), which shows some striking differences between maps created by 2D and 3D methods.

Figure 9 shows a comparison between maps made on the Groningen gas field for 2D and 3D surveys. The detail in the mapping of faults is quite striking and superior in the case of 3D.

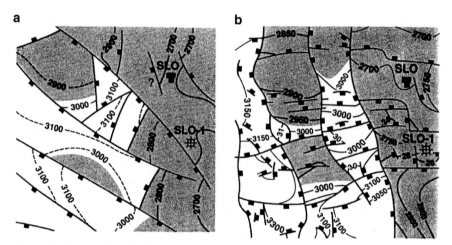

**Figure 9.** Nestvold's (1992) comparison of maps for Groningen field, based on (a) an interpretation of 2D surveys and (b) an interpretation of 3D surveys.

# Conclusions

With the success and popularity of 3D surveys, are there ever cases when we would not do 3D? The answer depends on cost — 3D surveys are more expensive than sets of 2D lines. For the most part, we will widely spaced 2D lines as reconnaissance surveys. If an interesting anomaly is found, we embark on a program of detailed seismic surveys, with 3D seismic surveys a major part of the effort, along with a possible wildcat well. The seismic and drilling activity generally are dictated by economics and the land situation.

As pointed out by Nestvold (1992), there is a growing industry consensus that 3D seismic surveys have a high economic benefit for most exploration plays. Certainly, 3D seismic surveys have become the main tool for reservoir characterization, as well as a major tool for field appraisal and for providing information for field development.

# References

Brown, A., 1991, Interpretation of three-dimensional seismic data: AAPG Memoir 42.

Cooper, D., 1995, 3-D seismic acquisition, design and quality control: CSEG course notes.

Cordsen, A., M. Galbraith, and J. Peirce, 2000, Planning land 3D seismic surveys: SEG.

French, W. S., 1974, Two-dimensional and three-dimensional migration of model-experiment reflection profiles: Geophysics, **39**, 265–277.

Liner, C. L., 1999, Elements of 3D seismology: PennWell Publishing.

Nestvold, E. O., 1992, 3D seismic: Is the promise fulfilled?: The Leading Edge, **11**, no. 6, 12–21.

Ratcliff, D., C. Jacewitz, and S. H. Gray, 1994, Subsalt imaging via target-oriented 3D prestack depth migration: The Leading Edge, **13**, no. 3, 163–170.

Stone, D. G., 1994, Designing seismic surveys in two and three dimensions: SEG.

Wu, W., L. R. Lines, and H. Lu, 1996, Analysis of higher-order, finite-difference schemes in 3D reverse-time migration: Geophysics, **61**, 845–856.

# Chapter 19

# Introduction to AVO Methods

**Note:** This chapter is derived largely from an unpublished report by Andrew Royle, which subsequently was condensed and revised by Laurence R. Lines (Royle, A., and L. R. Lines, 2003, Introduction to AVO methods: Unpublished course notes, University of Calgary).

## Introduction

In exploration seismology, the seismic-reflection method is used to find structures that have the potential to trap hydrocarbons. The risk lies in the possibility that the trap may contain no hydrocarbons. Exploration seismology would be more effective if the hydrocarbons could be distinguished directly on seismic sections.

In the 1960s, geophysicists discovered that the presence of gas is sometimes associated with the presence of high-amplitude reflections known as "bright spots" (Allen and Peddy, 1993). Bright spots are termed *direct hydrocarbon indicators* (DHIs). The use of bright spots for exploration greatly increased the success rate for wildcat gas wells. However, the bright-spot method has limitations, because lithologic conditions other than gas can cause bright reflections. Dry holes drilled on bright spots have found wet sands, lignites, carbonate or hard streaks, and igneous intrusions (Allen and Peddy, 1993).

Consequently, a more definitive indicator than bright spots on a stacked section is sought for direct detection of gas on seismic records. Ostrander (1984) demonstrated that gas-sand reflection coefficients differ anomalously with increasing offset and explained how to use this anomalous behavior as a direct hydrocarbon indicator on actual seismic data (Castagna and Backus, 1993).

We will discuss this method, which is called analysis of amplitude variation with offset (AVO).

## AVO, rock properties, and pore fluids

Seismic-reflection data are associated directly with subsurface rock properties. The Lamé parameters $\lambda$, $\mu$, and $\rho$ (which represent "incompressibility," rigidity, and density, respectively) can allow for enhanced identification of reservoir zones. This is partly because the compressibility of a rock unit is very sensitive to pore-fluid content. In addition, lithologic variations tend to be characterized better by fundamental changes in rigidity, incompressibility, and density rather than by changes in P-wave ($V_P$) and S-wave ($V_S$) velocities.

Figure 1a shows a rock matrix that is unstressed; the rock will have the maximum amount of pore space between grains. When compression (hydrostatic stress) is applied to a rock (Figure 1b), it squeezes the grains, causing a decrease in pore space. If a fluid such as oil or water is introduced to the pore space, it will resist compression by increasing the pressure against the grains, producing a more incompressible rock.

The introduction of gas into the pore space will lower incompressibility. Carbonates and igneous rocks have a harder framework and therefore have high incompressibility values, regardless of pore-fluid content.

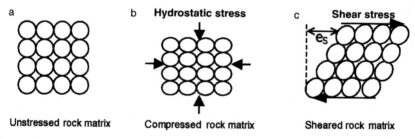

**Figure 1.** Comparison of rock matrix states for different stress states.

Figure 1c shows a sheared rock matrix in which shearing (shear stress) attempts to slide grains across one another. The pore-space volume remains essentially unchanged during shearing, regardless of pore-fluid type. Therefore, rigidity, which measures a rock's resistance to shearing, should be better able to characterize lithology than pore fluid. Shales are more susceptible to shearing than sands because of the nature of their grain orientation. Thus, they tend to exhibit low rigidity values. Carbonates, because of their more rigid framework, resist shearing and therefore have high rigidity values.

These elastic moduli are related to seismic-wave velocities by the following equations:

$$V_P^2 = \left( \frac{k+(4/3)\mu}{\rho} \right) = \frac{(\lambda+2\mu)}{\rho}, \tag{1}$$

where k is the bulk modulus, and

$$V_S^2 = \frac{\mu}{\rho}. \tag{2}$$

Goodway et al. (1997) emphasized these relationships between elastic moduli and elastic-wave velocities.

Investigation of the relative amplitudes within a CMP gather is known as amplitude variation with offset (AVO). Alternatively, if the relative amplitudes are examined with variation in reflection angle, the process is known as amplitude variation with angle (AVA). Reflections from gas-bearing rocks can show increasing amplitudes with increasing offset. However, for most reservoirs, an increase in amplitude is rare, and the majority of reflections observed on a CMP gather decreases in amplitude with offset. Therefore, AVO analysis is a search for an anomalous seismic response (Allen and Peddy, 1993).

In AVO analysis, we examine reflections at a range of source-receiver offsets. CMP gathers are the norm for modern seismic acquisition. Therefore, the input for AVO analysis imposes minimal or no additional effort in acquisition for most exploration. During seismic processing, traces are formed into a CMP gather, providing input for AVO analysis.

The use of AVO as a direct hydrocarbon indicator in clastic rocks is based on differences in the seismic-amplitude response, which is caused by the introduction of gas into pore spaces (Allen and Peddy, 1993). The theory of AVO for clastic rocks is quite simple. When a P-wave strikes a rock interface at nonnormal incidence, a fraction of the incident P-wave energy is converted to S-wave energy. The contrast in the P-wave velocity and S-wave velocity causes different reflection responses for the case of a gas-sand/shale contact as compared to a gas-sand/wet-sand contact.

An important rock-property parameter, which varies as a function of P-wave and S-wave velocity ratio, is Poisson's ratio. Poisson's ratio has a physical definition. If one takes a cylindrical rod of an isotropic elastic material and applies a small axial compressional force to the ends, the rod will change shape. The length of the rod will decrease slightly and the radius will increase slightly (Ostrander, 1984). By definition, Poisson's ratio is the ratio of the relative change in radius (fractional transverse contraction) to the relative change in length (longitudinal extension) when the rod is stretched (Sheriff, 1991).

Materials generally have Poisson's ratios of 0.0–0.5. Materials such as fluids, which have a shear modulus (rigidity) equal to 0.0, have a Poisson's ratio of 0.5. Gas sands have ratios closer to 0.0. Poisson's ratio can be established by using field or laboratory measurements of P-wave and S-wave velocities.

Poisson's ratio for an isotropic elastic material is simply related to the P-wave ($V_P$) and S-wave ($V_S$) velocities of the material by:

$$\sigma = \frac{(V_P/V_S)^2 - 2}{2[(V_P/V_S)^2 - 1]}. \tag{3}$$

In AVO analysis, it is important to note that P-waves and S-waves have quite different sensitivities to pore fluids. The introduction of a small amount of gas into the pore spaces of a clastic sedimentary rock can reduce the P-wave velocity of the rock drastically, but the S-wave velocity is relatively unaffected. Because S-waves cannot travel through liquids or gases, the pore-space constituent does not affect the velocity of S-waves, whose velocity depends directly on the rock framework (matrix). In fact, the S-wave velocity may increase with the introduction of gas into the pore space because the density of gas is lower than the density of brine. Figure 2 shows the theoretical values for S-wave and P-wave velocities at various gas saturations for a porous medium.

AVO has been used widely in clastic reservoirs as a direct hydrocarbon indicator. On the other hand, carbonate reservoirs are different because of their rigid framework. AVO has been used in carbonate reservoirs in an attempt to detect porosity. Porosity has a major effect on the acoustic impedance of a limestone. Seismic-reflection coefficients from the top of limestone beds are sometimes lower where porosity is high (dimming). Examples of amplitude dimming were shown in Chapter 13 of this book, on carbonate reefs. These amplitude effects can be detected on stacked seismic data, but we cannot distinguish porosity from facies changes. Offset-dependent effects, which are a function of additional physical properties of rocks, possibly could distinguish between tight limestone facies and porous limestone facies.

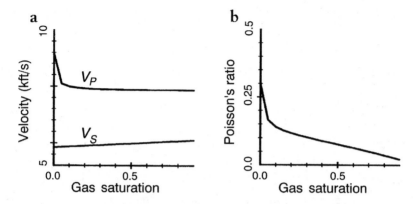

**Figure 2.** (a) P-wave velocity of a porous solid rapidly decreases with the introduction of a small percentage of gas. The S-wave velocity increases linearly with gas saturation. (b) The two effects show a decrease in the Poisson's ratio of the rock with increasing gas saturation (Allen and Peddy, 1993).

In carbonate reservoirs, AVO has not proved itself as an identifier of pore fluid but possibly can identify porosity development. The first step in applying AVO would be to use seismic modeling, then compare those synthetic traces to the actual seismic data for evaluation. The amplitude behavior of the synthetic traces is found to be quite sensitive to the $V_P/V_S$ ratio specified in the reservoir zone of the earth model.

For example, Chacko (1989) used AVO successfully in an attempt to detect porosity by comparing AVO synthetic seismograms to the corresponding CMP gathers of the seismic data. Chacko (1989) showed that gas-charged porous limestone facies might have lower $V_P/V_S$ values than wet porous-limestone facies. The application of AVO to carbonate reservoirs is still not an intuitive process, and the accuracy may depend on good well control and measured S-wave velocities in the area.

# Offset-dependent reflectivity and Zoeppritz's equations

## Compressional wave propagation

A plane P-wave striking an interface at normal incidence experiences no conversion to an S-wave at the interface. At any angle other than normal incidence, some fraction of the incident P-wave is partially converted to an

S-wave at the interface, and the reflection coefficients become a function of $V_P$, $V_S$, and density ($\rho$) of each layer (Figure 3). In Figure 3, we consider the reflected and transmitted arrivals that result from a P-wave that is incident on the boundary. Solutions to the boundary conditions involve energy partitioning from an incident P-wave to a reflected P-wave, a reflected S-wave, a transmitted P-wave, and a transmitted S-wave.

The angles for incident, reflected, and transmitted rays at the boundary are related by Snell's law. When a wave crosses a boundary between two isotropic media, the wave changes direction such that the ray parameter, $p$, stays constant. This is an expression of Snell's law. If we denote P-wave angles by $\theta$ and S-wave angles by $\phi$, we have

$$p = \frac{\sin \theta_i}{V_{P1}} = \frac{\sin \theta_r}{V_{P1}} = \frac{\sin \theta_t}{V_{P2}} = \frac{\sin \phi_r}{V_{S1}} = \frac{\sin \phi_t}{V_{S2}}. \tag{4}$$

Here, $\theta_i$ is the angle of the incident P-wave. $\theta_r$ and $\phi_r$ are the angles of reflection of the P- and S-waves in medium 1, which have velocities $V_{P1}$ and $V_{S1}$, respectively. The angles $\theta_t$ and $\phi_t$ are the transmission angles of the P- and S-waves in medium 2, which have velocities $V_{P2}$ and $V_{S2}$ (Sheriff, 1991).

**Figure 3.** An incident P-wave is split into four components upon interacting with the boundary between two media.

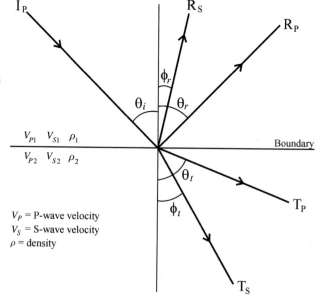

$V_P$ = P-wave velocity
$V_S$ = S-wave velocity
$\rho$ = density

## Reflection coefficients

The P-wave reflection coefficient as a function of incidence angle $R_{PP}(\theta_i)$ is defined as the ratio of the amplitude of the reflected P-wave to that of the incident P-wave. In addition, the P-wave transmission coefficient $T_{PP}(\theta_i)$ is the ratio of amplitude of the transmitted P-wave to that of the incident P-wave (Castagna and Backus, 1993). The reflection and transmission of elastic plane waves at a boundary can be described in terms of amplitudes, displacement potentials, and energy (Young and Braile, 1976).

From the acoustic impedance values, we can compute the reflection coefficients. For the special case of normal-incidence reflections from an interface that separates media of densities $\rho_1$ and $\rho_2$ and velocities $V_1$ and $V_2$, the reflection coefficient for a plane wave is

$$R = \frac{V_2\rho_2 - V_1\rho_1}{V_2\rho_2 + V_1\rho_1}. \qquad (5)$$

The reflection coefficient gives the relative amplitudes of the reflected and incident waves. We note that when a P-wave travels from low impedance to high impedance, there is no phase shift, but when it travels from high to low impedance, there is a 180° phase change.

The transmission coefficient relates the size of the transmitted amplitude to the incident amplitude. The P-wave transmission coefficient at normal incidence is given by the equation

$$T_p = 1 - R_p. \qquad (6)$$

The variation of reflection and transmission coefficients with angles of incidence is referred to as offset-dependent reflectivity and is the fundamental basis for AVO analysis (Castagna and Backus, 1993).

## AVO and Zoeppritz's equations

The previous equations for reflection and transmission coefficients in equations 5 and 6 are for the cases of normal incidence. With AVO, we consider the reflected amplitudes at varying angles of incidence. For this analysis, we generally consider some form of Zoeppritz's equations.

Zoeppritz's equations describe a reflection from an interface that separates two isotropic elastic media with different values for $V_P$, $V_S$, and $\rho$. For an incident plane wave on the interface, the equations describe the reflected and transmitted P-waves and S-waves. This requires two fundamental sets of boundary conditions. First, we require continuity of displacement, which states that there is no slipping along an interface or no ripping apart of rocks at the interface. Second, we require the continuity of traction (projection of the stress tensor). In 2D models, this results in four equations with four unknowns, and the solutions can be derived in terms of $R_P$, $T_P$, $R_S$, and $T_S$.

Zoeppritz's equations give the reflection and transmission coefficients for plane waves as a function of angle of incidence and six independent elastic parameters, three on each side of the reflecting interface (Shuey, 1985). These equations are quite complex and are not sufficiently straightforward to fit to actual seismic data. Because the solutions of Zoeppritz's equations are complex, an intuitive grasp of the result for different cases is difficult. Generally speaking, AVO analysis deals with approximations to Zoeppritz's equations.

## AVO methodology approximations to Zoeppritz's equations

Approximations to Zoeppritz's equations have been developed by many researchers. Aki and Richards (1980), Shuey (1985), Hilterman (1990), Smith and Gidlow (1987), and Fatti et al. (1994) have simplified the relationship between reflection coefficient and angle of incidence so that the major factors could be identified. We will briefly discuss the approximations of these researchers, which allow AVO analysis to be applied without difficulty.

Aki and Richards (1980) gave Zoeppritz's equations in a matrix form:

$$Q = P^{-1}R. \tag{7}$$

(This analysis is discussed in more detail in Appendix A of this chapter.) $Q$ is defined as the matrix of reflected arrival amplitudes, and $P$, $R$ contain coefficients dependent on reflection amplitudes. The Aki and Richards approximation gave a simpler and more practical form of these equations that can be applied more readily to real seismic data. The Aki and Richards approximation is given by

$$R(\theta) = \frac{1}{2}\left(1 - 4V_S^2 p^2\right)\frac{\Delta\rho}{\rho} + \frac{1}{2\cos^2\theta}\frac{\Delta V_P}{V_P} - 4V_S^2 p^2 \frac{\Delta V_S}{V_S}, \tag{8}$$

where $R$ = offset dependent reflectivity, $\Delta V_P = (V_{P2} - V_{P1})$, $V_P$ = P-wave velocity $(V_{P2} + V_{P1}) / 2$, $\Delta V_S = (V_{S2} - V_{S1})$, $V_S$ = S-wave velocity $(V_{S2} + V_{S1})$

/ 2, $\rho$ = density average $(\rho_2 + \rho_1) / 2$, $\Delta\rho = (\rho_2 - \rho_1)$, $p$ = raypath parameter, and $\theta$ = average angle of incidence.

The Aki and Richards (1980) approximation makes the following assumptions: (1) The relative changes of property are sufficiently small (as discussed later), (2) second-order terms can be neglected, and (3) the incident angle does not approach the critical angle. The Aki and Richards approximation is sufficient for most acquired seismic data. However, there are several approximations that further simplify the Aki and Richards approximation. These later approximations are more tractable and yielded meaningful geologic insights into interpretation.

Shuey's (1985) approximation of Zoeppritz's equations reparameterized the Aki and Richards approximations in terms of impedance and Poisson's ratio change, $\Delta\sigma$, which is the difference in Poisson's ratio $(\sigma_2 - \sigma_1)$ between two layers. The Shuey-type approximation commonly is used in AVO because it expresses the reflection coefficient as a sum of three terms containing a normal-angle term, a near-angle term, and a far-angle term. A complete description of these approximations is given by Allen and Peddy (1993).

Shuey's approximation gives a relatively simple relationship between rock properties (Poisson's ratio) and the variation in reflection coefficients, and this stresses the importance of Poisson's ratio as the primary determinant of the AVO response of a reflection.

Shuey's approximation can be simplified even further to a form given by

$$R(\theta) = R_0 + G\sin^2\theta \,, \tag{9}$$

where $R_0$ is the normal incident P-wave reflectivity, or "intercept," and $G$ is the "gradient" term. The gradient, by definition, is the rate of change of the amplitudes at each time sample as a function of incidence angle on a CDP gather. The gradient should contain the entire AVO effect. The intercept represents the theoretical zero-offset response. This response will show bright spots but does not show any AVO effect. The intercept and gradient terms from this approximation can be obtained easily through linear regression.

Hilterman (1989) introduced an approximation of Shuey's (1985) equation which assumed that $V_S/V_P = 0.5$ or $\sigma = 0.33$, and that $tan^2\theta \cong sin^2\theta$. Hilterman's equation is given by

$$R(\theta) = NI\cos^2\theta + \left[\frac{\Delta\sigma}{(1-\sigma)^2}\right]\sin^2\theta \,, \tag{10}$$

where $NI$ = normal-incidence reflection coefficient, $\Delta\sigma$ = the difference of

Poisson's ratio between the lower and upper media, and $\theta$ = angle of incidence.

Hilterman refers to the term $[\Delta\sigma/(1-\sigma)^2]$ as the Poisson reflectivity (*PR*). This term is quite similar to the gradient term from the Shuey approximation. Furthermore, because the average Poisson's ratio is set to 0.33, equation 10 simplifies to

$$R(\theta) = NI \cos^2 \theta + 2.25\Delta\sigma \sin^2 \theta. \tag{11}$$

The "normal-incidence" and "Poisson-reflectivity" terms from this approximation can be estimated through linear regression by fitting this model to the observed AVO data. These estimated values of the normal incidence and Poisson's reflectivity then could be used to predict lithologies, as described by Hilterman (1990).

Although Hilterman's equation is an approximation, it allows us to readily understand AVO effects. For example, let us consider the two shale-sand situations that are important to distinguish, from an exploration point of view.

In one example, we consider a shale overlying a brine-saturated sandstone. Let the shale have a density of 2.2 gm/c$^3$, a P-wave velocity of 3000 m/s, and a Poisson's ratio of 0.33. This shale overlies a brine-saturated sandstone with a density of 2.2 gm/c$^3$, a P-wave velocity of 2500 m/s, and a Poisson's ratio of 0.33. If we substitute into equation 11, we obtain

$$R(\theta) = -0.09 \cos^2 \theta. \tag{12}$$

The reflection coefficient for this situation has a small negative value at normal incidence, which will decrease to 0 as the reflection angle increases.

In a second example, we consider the same shale overlying a gas-saturated sandstone, with a lower Poisson's ratio than in the previous example. Consider the sandstone to have the same density and P-wave velocity as the previous example but a Poisson's ratio of 0.10, giving a Poisson's ratio contrast. Therefore, the reflection coefficient as a function of angle in this example becomes

$$R(\theta) = -0.09 \cos^2 \theta - 0.52 \sin^2 \theta. \tag{13}$$

The reflection coefficient again has a small negative value at normal incidence, but it becomes more negative as the reflection angle increases. In the gas-saturated case, the magnitude of the reflection will increase with offset.

This is the classic "Class 3 sand," as outlined by Rutherford and Williams (1989) in their AVO classification of gas sands.

The "fluid-factor" concept introduced by Smith and Gidlow (1987) attempts to highlight gas-bearing sandstones. The starting point for Smith and Gidlow's (1987) approach is a linearized approximate form of Zoeppritz's equations after Aki and Richards (1980), in which $V_S/V_P$ is determined by using empirical relations.

Because the values of $V_S$ and $V_P$ are not estimated easily from seismic reflectivities alone, it is advisable to make some assumption about the $V_S/V_P$ ratio. To determine $V_S/V_P$, we can use the empirically derived "mud-rockline" relationship between $V_P$ and $V_S$ for water-saturated clastic rocks, as proposed by Castagna et al. (1985):

$$V_P = 1369 + 1.16V_S. \tag{14}$$

Equation 14 is a "global" relationship, and a more regional trend may be appropriate for specific areas. Using Castagna's mud-rock relationship, Smith and Gidlow derived the fluid factor in terms of the P-wave and S-wave reflectivities given by

$$\Delta F = R_P - 1.16\left(\frac{V_S}{V_P}\right)R_S, \tag{15}$$

where $R_P$ = zero-offset P-wave reflection coefficient and $R_S$ = zero-offset S-wave reflective coefficient.

The fluid factor will have a low magnitude for all reflectors in a clastic sedimentary sequence except for rocks that lie off the mud-rock line. The fluid factor will have a high value for gas-saturated sandstones, thereby providing one of the most useful tools in AVO analysis.

# Pitfalls and possibilities in AVO analysis

AVO analysis has many pitfalls. In other words, there are many ways to perform AVO incorrectly. First, we require "true amplitudes" in our input seismic data. This is important, because seismic amplitudes can be distorted by many stages of processing, including steps such as gain application, trace mixing, deconvolution, and migration. Second, even if amplitudes are perfect, there are ambiguities in estimating rock properties from seismic amplitudes, as outlined by Downton and Lines (2001). Moreover, it is important to understand the approximations in our AVO models, as outlined in this chapter, and to know whether these approximations are appropriate.

Despite these difficulties, there are several worthwhile applications of AVO, as outlined in books by Allen and Peddy (1993) and by Castagna and Backus (1993). Most of these applications involve the detection of gas in sandstones, but there are also possibilities with carbonate plays.

## Appendix A

The following discussion is a more detailed explanation of Aki and Richards' (1980) approximation of Zoeppritz's equations.

Aki and Richards (1980) expressed Zoeppritz's equations in a convenient matrix arrangement. For an interface between two elastic half-spaces, there are 16 reflection and transmission coefficients (Castagna and Backus, 1993). Aki and Richards use a special notation to denote the type of incident wave and the type of derived wave. Figure A-1 explains their notation.

From this notation, the scattering matrix is given by

$$
Q = \begin{bmatrix}
\overset{\backslash\;/}{PP} & \overset{\backslash\;/}{PS} & \overset{\backslash\;\backslash}{PP} & \overset{\backslash\;\backslash}{PS} \\
\overset{\backslash\;/}{SP} & \overset{\backslash\;/}{SS} & \overset{\backslash\;\backslash}{SP} & \overset{\backslash\;\backslash}{SS} \\
\overset{/\;/}{PP} & \overset{/\;/}{PS} & \overset{/\;\backslash}{PP} & \overset{/\;\backslash}{PS} \\
\overset{/\;/}{SP} & \overset{/\;/}{SS} & \overset{/\;\backslash}{SP} & \overset{/\;\backslash}{SS}
\end{bmatrix} = \underset{\sim}{P}^{-1}\underset{\sim}{R}
$$

where

$$
\underset{\sim}{P} = \begin{bmatrix}
-\sin\theta_1 & -\cos\phi_1 & \sin\theta_2 & \cos\phi_2 \\
\cos\theta_1 & -\sin\phi_1 & \cos\theta_2 & -\sin\phi_2 \\
2\rho_1 V_{S1}\sin\phi_1\cos\theta_1 & \rho_1 V_{S1}\left(1-2\sin^2\phi_1\right) & 2\rho_2 V_{S2}\sin\phi_2\cos\theta_2 & \rho_2 V_{S2}\left(1-2\sin^2\phi_2\right) \\
-\rho_1 V_{P1}\left(1-2\sin^2\phi_1\right) & \rho_1 V_{S1}\sin 2\phi_1 & \rho_2 V_{P2}\left(1-2\sin^2\phi_2\right) & -\rho_2 V_{S2}\sin 2\phi_2
\end{bmatrix}
$$

and

$$
R = \begin{bmatrix}
\sin\theta_1 & \cos\phi_1 \\
\cos\theta_1 & -\sin\phi_1 \\
2\rho_1 V_{S1}\sin\phi_1\cos\theta_1 & \rho_1 V_{S1}\left(1-2\sin^2\phi_1\right) \\
\rho_1 V_{P1}\left(1-2\sin^2\phi_1\right) & -\rho_1 V_{S1}\sin 2\phi_1
\end{bmatrix}
$$

$$
\begin{bmatrix}
-\sin\theta_2 & -\cos\phi_2 \\
\cos\theta_2 & -\sin\phi_2 \\
2\rho_2 V_{S2}\sin\phi_2\cos\theta_2 & \rho_2 V_{S2}\left(1-2\sin^2\phi_2\right) \\
-\rho_2 V_{P2}\left(1-2\sin^2\phi_2\right) & \rho_2 V_{S2}\sin 2\phi_2
\end{bmatrix}.
$$

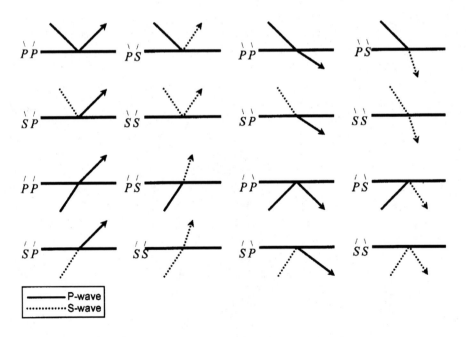

**Figure A-1.** Notation for the 16 possible reflection/transmission coefficients.

# References

Aki, K., and P. G. Richards, 1980, Quantitative seismology: W. H. Freeman & Co.

Allen, J. L., and C. P. Peddy, 1993, Amplitude variation with offset — Gulf Coast case studies: SEG.

Castagna, J. P., and M. M. Backus, eds., 1993, Offset-dependent reflectivity — Theory and practice of AVO analysis: SEG.

Castagna, J. P., M. L. Batzle, and R. L. Eastwood, 1985, Relationships between compressional-wave and shear-wave velocities in clastic silicate rocks: Geophysics, **50**, 571–581.

Chacko, S., 1989, Porosity identification using amplitude variations with offset: Examples from South Sumatra: Geophysics, **54**, 942–951.

Downton, J. E., and L. R. Lines, 2001, Constrained three-parameter AVO inversion and uncertainty analysis: 71st Annual International Meeting, SEG, Expanded Abstracts, 251–254.

Fatti, J. L., G. C. Smith, P. J. Vail, P. J. Strauss, and P. R. Levitt, 1994, Detection of gas in sandstone reservoirs using AVO analysis: Geophysics, **59**, 1362–1376.

Goodway, B., T. Chen, and J. E. Downton, 1997, Improved AVO fluid detection and lithology discrimination using Lamé petrophysical parameters: "$\lambda\rho$," "$\mu\rho$," and "$\lambda/\rho$ fluid stack," from *P*- and *S*- inversions: Canadian Society of Exploration Geophysicists Recorder, **22**, 3–5.

Hilterman, F. J., 1989, Is AVO the seismic signature of rock properties?: 59th Annual International Meeting: SEG, Expanded Abstracts, 559.

Hilterman, F. J., 1990, Is AVO the seismic signature of lithology? A case history of the Ship Shoal–South addition: The Leading Edge, **9**, no. 6, 15–22.

Ostrander, W. J., 1984, Plane wave reflection coefficients for gas sands at nonnormal angles of incidence: Geophysics, **49**, 1637–1648.

Royle, A., and L. R. Lines, 2003, Introduction to AVO methods: Unpublished course notes, University of Calgary.

Rutherford, S. R., and R. H. Williams, 1989, Amplitude-versus-offset variations in gas sands: Geophysics, **54**, 680–688.

Sheriff, R., 1991, Encyclopedic dictionary of exploration geophysics: SEG.

Shuey, R. T., 1985, A simplification of the Zoeppritz equations: Geophysics, **50**, 609–614.

Smith, G. C., and P. M. Gidlow, 1987, Weighted stacking for rock property estimation in gas sands: Geophysical Prospecting, **35**, 993–1014.

Young, G. B., and L. W. Braile, 1976, A computer program for the application of Zoeppritz amplitude equations: Bulletin of the Seismological Society of America, **52**, 923–956.

# Chapter 20

# Reservoir Characterization

Several factors will cause oil and gas producers to focus increasingly on characterization of petroleum reservoirs in the near future. Experts predict that more than 95% of the world's oil production in the 21st century will come from existing fields. Recovery rates from present producing fields are often quite modest. Much less than 50% of available oil usually is produced from a reservoir. Moreover, there is still an insatiable thirst for oil and gas consumption among the world's population. Petroleum is also necessary for the production of many chemicals and plastics. All these factors impact the need for enhanced recovery of petroleum. Increased production will be made possible only through effective reservoir characterization.

Reservoir characterization is a multidisciplinary field that attempts to describe petroleum deposits and the nature of the rocks that contain hydrocarbons. Reservoir characterization relies on expertise from petroleum engineering, geology, and geophysics. The integration of information from these fields focuses on various aspects of the reservoir. The formation of reservoir-characterization teams requires the synergistic teamwork of engineers, geologists, and geophysicists. Books that describe this emerging science include those by Sheriff (1992) and by Gadallah (1994).

The fundamental goals of reservoir characterization are to determine the following:

- presence of hydrocarbons
- reservoir porosity
- reservoir permeability

Many attempts have been made to detect the presence of hydrocarbons with geophysical methods. These direct hydrocarbon indicators (DHIs) are related largely to seismic amplitudes. The analysis of seismic "bright spots," popular in the 1970s, relied on interpretation of stacked data. During the 1980s and 1990s, an increased emphasis was placed on AVO analysis of prestack data. Chapter 19 of this book describes the many applications of AVO and its relation to lithology and the presence of reservoir fluids.

# Rock physics

Rock physics is a key component of reservoir analysis, and it provides crucial information about the source rock, the seals, and the capacity of the reservoir to contain hydrocarbons. Porosity is a key ingredient, and it will determine the supply of petroleum contained in the rocks. Seismic velocity can be related to porosity. Wyllie's time-average equation relates velocity to porosity by using the time average of seismic travel through the rock matrix and the fluid-filled pores. Wyllie's equation is reasonably valid for sandstones but is an oversimplification for carbonates. As a rule, velocity decreases will accompany porosity increases.

One of the biggest challenges in reservoir characterization is the estimation of permeability, or the flow of fluids through the reservoir. To some degree, permeability is related to the interconnection of pores; hence, it is often related to porosity. In addition, permeability is influenced by fractures and fracture direction. We will explore fracture permeability further in our treatment of shear waves and multicomponent data, in Chapter 22. Seismic data usually are affected by permeability, but the connection is often difficult.

# Integrated reservoir characterization — A scaling problem

Several sources of information help to characterize the physical properties of the reservoir. Various geologic and geophysical data measure different reservoir volumes at different scales. It is the purpose of reservoir characterization to integrate these data sets.

The primary tool for probing the entire reservoir volume is 3D-reflection seismic surveying. Although 3D seismic data should cover the entire reservoir, the wavelengths of surface seismic data (given by velocity/frequency) generally are 10–100 m, because most seismic frequencies for surface recording are in the range of 5–100 Hz.

Well logs and core measurements usually are made on the scale of 0.1–1.0 m. Although the scale of such measurements in quite fine, the volume of investigation is very small — within a few meters of the wellbore. To fill in the "missing wavelengths" from 1–10 m over significant volumes of the reservoir, we acquire borehole seismic information. These data are gathered by using sources and receivers in boreholes to fill the wavelength gap. We already have discussed these methods, which include vertical seismic profiling and cross-borehole surveys.

Many of the insights in the area of reservoir sampling have been explained by Dr. Jerry Harris and his colleagues at Stanford University. Table 1 gives a comparison of the data types and scales. Harris et al. (1995) advocated the filling of the missing wavelengths by use of borehole data.

**Table 1.** A comparison of data types of reservoir sampling.

| Data type | Reservoir volume | Wavelength of investigation |
|---|---|---|
| 3D surface seismic data | Entire reservoir volume | 10–100 m (dependent on velocity and frequency) |
| Borehole seismic data | Areas (volumes) between well and sources | VSP 5–50 m Cross-borehole seismology 1–10 m |
| Well data | Volume of rock near the wellbore | 10 cm–1 m |

Figure 1 shows the wavelengths of various surface and borehole measurements. Surface wavelengths are the coarsest (although they span the entire reservoir volume). The crosswell velocity tomograms and reflection images show that the resolution in crosswell data is an order of magnitude greater than in surface seismic data. If we superimpose the velocity obtained from crosswell measurements, we see that it is similar to a low-frequency version of the sonic data (although the crosswell data sample the rocks between wells, whereas logs are at the well site). The logs and cores show a high-frequency but low-volume coverage between wells. In essence, all the reservoir information is useful. We need to integrate all data into a description of the reservoir.

A case-history example of integrated reservoir characterization is shown in a paper by Lines et al. (1995) on the North Cowden field in west

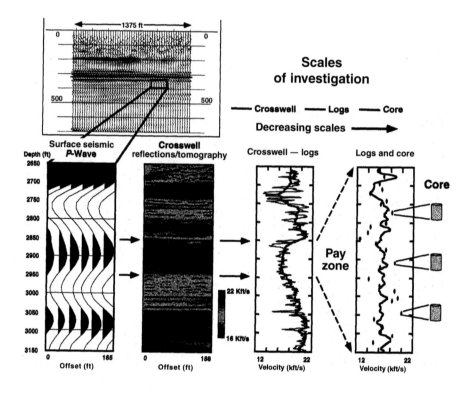

**Figure 1.** Comparison of wavelength scales for surface seismic data, crosswell images (both velocity and reflectivity), crosswell velocities, and sonic logs, as well as logs and cores (from Harris et al., 1995).

Texas. The example shown in Figure 2 compares the crosswell reflection image with a corresponding section of the 3D seismic survey. The crosswell image acts as a "seismic magnifying glass" for the larger 3D survey. This case study attempts to integrate these data, as well as VSPs, logs, and cores, to describe the porous siltstone layer of the Grayburg unit.

## Conclusions

Reservoir characterization is the art and science of integrating different data types. It attempts to integrate geologic and geophysical data on different scales to form a picture of the reservoir and to integrate the knowledge of the reservoir engineer, geologist, and geophysicist. The 21st century will see many exciting developments in reservoir characterization.

**a**

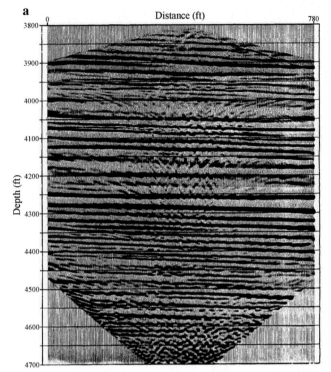

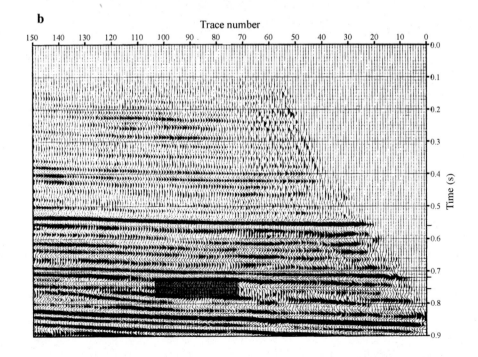

**b**

**Figure 2.** Comparison of the crosswell reflection image (a) with the surface seismic reflection line (b) of the North Cowden field in west Texas. The shaded area on the surface data indicates the corresponding area for the crosswell survey (from Lines et al., 1995).

# References

Gadallah, M., 1994, Reservoir seismology: PennWell Books.

Harris, J. M., R. Nolen-Hoeksema, R. T. Langan, M. Van Schook, S. Lazarotos, and J. W. Rector, 1995, High-resolution crosswell imaging of a west Texas carbonate: Geophysics, **60**, 667–682.

Lines, L. R., et al., 1995, Integrated reservoir characterization: Beyond tomography: Geophysics, **60**, 354–364.

Sheriff, R., 1992, Reservoir geophysics: SEG.

# Chapter 21

# Time-lapse Seismology

In our previous discussions, we described petroleum reservoirs as having 3D variation in space but no variation in the fourth dimension, time. Although the assumption of time invariance is a suitable for geologic processes, it is not suitable for enhanced oil recovery (EOR), because EOR will create physical changes to the reservoir within weeks or months.

## Introduction

It is important for reservoir engineers and geoscientists to monitor such reservoir changes by repeated seismic experiments, termed *time-lapse seismology* (sometimes termed *4D seismology*). The most widespread use of time-lapse seismology experiments has involved the monitoring of steam injection into heavy-oil reservoirs. Significant decreases of P-wave velocity (on the order of 30%) occur when oil sands are heated from temperatures of about 25°C to temperatures of 140°C, as demonstrated originally by Nur (1982). This sudden change in velocity, with concomitant heating of the oil sands, means that seismic responses will change significantly as a result of steam injection. This allows for the mapping of steam fronts or heated zones as a function of time.

Figure 1 shows a map of how seismic velocities decreased near wells of steam injection. This map is the result of steam injection into the Athabasca oil sands near Gregoire Lake, Alberta, Canada.

Time-lapse seismology has had widespread application in the Alberta oil-sands projects. This is economically very important, because the Athabasca, Cold Lake, Lloydminster, and Wabasca oil-sands deposits of northern Alberta have reserves of about one trillion barrels of oil.

**Figure 1.** Map of seismic-velocity decrease caused by steam injection at wells marked by white circles (from Lines et al., 1990). Velocity units are in km/s. Velocities prior to steam injection were 2.40 km/s.

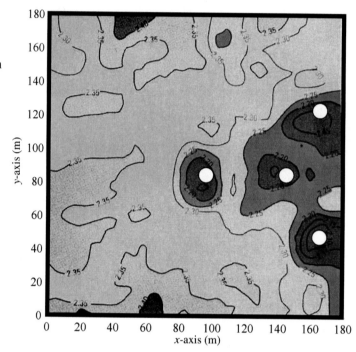

# Seismic monitoring of steam zones — A case study from Pikes Peak, Saskatchewan

Various seismic techniques can be used for monitoring zones of steam injection in heavy-oil recovery. The following case study is an abbreviated version of a paper by Watson et al. (2002). In this example, various techniques are used to delineate the steamed reservoir at Husky Energy's Pikes Peak heavy-oil field in Saskatchewan, Canada.

We compare the methods of reflectivity differencing, impedance differencing, P-wave traveltime ratios, and an isochron method for examining $V_P/V_S$ ratios. The differencing techniques use repeat seismic surveys and show the expected lowering of P-wave velocity in the zone of steam injection. The $V_P/V_S$ isochron method uses a single multicomponent survey and compares interpreted traveltimes around the reservoir for P-P (compressional) and P-S (converted) wave data. All four methods show promise for delineating areas of steam injection.

Steam-zone delineation by seismic methods has been applied successfully during the past two decades. Most methods are based on ideas originated by Nur (1982), who demonstrated that P-wave velocity is lowered significantly as temperature increases in heavy-oil sands. Nur's results have led to many time-lapse seismology projects in western Canada's heavy-oil fields. The applications of seismic monitoring for Athabasca oil sands were discussed by Pullin et al. (1987), de Buyl (1989), Lines et al. (1990), and Matthews (1992). Further advances for seismic monitoring of enhanced oil recovery at Cold Lake, Alberta, were made by Eastwood (1993) and by Eastwood et al. (1994).

The following case history describes seismic monitoring efforts at Pikes Peak field, just east of the Alberta-Saskatchewan border near Lloydminster. Husky Energy Ltd. has operated the heavy-oil field and has produced more than 42 million barrels of oil using steam-drive enhanced oil recovery (EOR). The effective viscosity of the oil is reduced, and the mobility is increased in the reservoir as steam is injected at high temperatures and pressures. The oil is produced either from neighboring wellbores or through the same well-bore used for cyclic injection.

The Lower Cretaceous Waseca Formation is the producing reservoir, at a depth of about 450 m. It is an incised valley filled with estuarine deposits of a basal, homogeneous sand unit, an interbedded sand and shale unit, and a capping shale unit (Van Hulten, 1984).

In this study, Watson et al. (2002) examined four techniques for seismic detection of steam fronts:

1) differencing of reflectivity functions for the monitor and base surveys

2) differencing of acoustic-impedance estimates for the monitor and base surveys

3) comparison of interval P-wave traveltimes for the monitor and base surveys

4) estimation of $V_P/V_S$ variation from multicomponent data

We compare and contrast the results of these approaches as a means of detecting steam fronts. A fifth method, outlined by Hedlin et al. (2002), describes the variation of Q (inverse-attenuation) with temperature, and this method also shows considerable promise.

Seismic, well log, and production data were provided by Husky Energy. Husky originally acquired a set of 2D seismic swaths in 1991. To investigate time-lapse effects, the University of Calgary, Alberta Oil Sands Technology Research Authority (AOSTRA), and Husky returned to the field in March 2000 to acquire a three-component line on the eastern side. During acquisition, four components were collected: P-wave (vertical and array), SV-wave, SH-wave, and experimental surface-microphone data.

The original and repeat seismic data were processed simultaneously using similar workflows. The most significant difference between the two surveys was the final bandwidth. The 1991 survey used a vibrator sweep of 14–110 Hz, and the 2000 survey used a bandwidth of 14–150 Hz. Therefore, time-lapse differencing required the second survey to be high-cut filtered to the 14–110 Hz range.

Both surveys were conducted in winter, possibly minimizing ground-coupling differences. Time-lapse seismology requires that the survey's acquisition and processing be as similar as possible so that seismic differences are the result of changes in reservoir properties.

Figures 2 and 3 show the reflectivity sections (vertical component) for H1991 and H2000 lines, respectively. The Waseca and Sparky reflectors are shown on both sections. Reflections beneath the Waseca, including the Devonian reflection at about 700 ms, are slightly later in time on the H2000 section. We note the higher-frequency reflectors for the H2000 line in Figure 3, caused by the higher vibroseis sweep.

Wells D15-06, C08-06, and D02-06 were used to create synthetic ties to the P-wave seismic data because they had original sonic and density logs over the Waseca interval. One well, 1A15-06 (which is about 120 m south of D15-06), was used to tie to the converted-wave (PS) seismic data because it had a dipole sonic log run in 2000, which allowed arrival times to be estimated for the P-wave and converted-wave sections. The seismic interpretation was constrained with these forward-modeled ties.

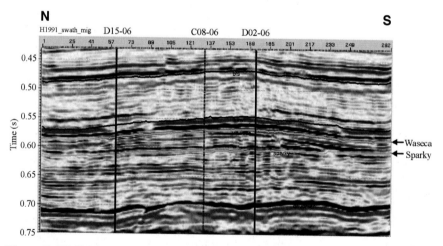

**Figure 2.** H1991 interpreted P-wave reflectivity section (from Watson et al., 2002).

The seismic-reflectivity data were used to invert for acoustic impedance (AI). Acoustic impedance is different from reflectivity data; it is an interval property rather than an interface property. Consider a simple two-layer model that has AI values of $Z_i$ and $Z_{i+1}$, where $Z_i = \rho_i V_i$ for a given interval, $i$. Equation 1 (McQuillin et al., 1984) derives the reflection coefficient, $R_i$, at an interface:

$$R_i = \frac{Z_{i+1} - Z_i}{Z_{i+1} + Z_i}.$$ (1)

Rearranging equation 1 to solve for $Z_{i+1}$ gives equation 2:

$$Z_{i+1} = \frac{Z_i(1 + R_i)}{1 - R_i}$$ (2)

Given a reflectivity sequence such as a seismic trace and an initial AI value, $Z_i$, the trace can be inverted to give acoustic impedance with traveltime or depth, as equation 2 is used iteratively. This method is described by Lindseth (1979).

We performed a type of "trace-based" inversion, in which information from nonseismic data sources constrains the results. The acoustic impedance from the inversion has the effect of removing wavelet sidelobe energy and tuning. This improves interpretation of boundaries and allows evaluation of the internal rock properties.

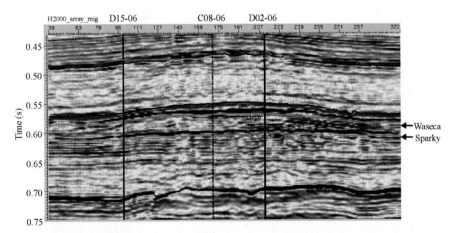

**Figure 3.** H2000 reflectivity section (note higher-frequency content) (from Watson et al., 2002).

Injecting steam into the reservoir formation reduces the AI as a result of lowering the seismic velocities. A consequence is an increased traveltime in the reservoir. Seismic data have been acquired over the field to determine the extent of the steam front.

An important aspect of the inversion process is the choice of wavelet. Because of the different frequency content of the two time-lapse lines, two different wavelets for the surveys were required. Wavelets were estimated by using the method described in Chapter 8 of this book. Wavelet estimation was an iterative process that improved as the synthetic tie and wavelet improved.

Prior to subtraction of the reflectivity and impedance sections, they needed to be compared and calibrated to adjust for static, amplitude, and phase differences. It was critical that the time window for the calibration be above the zone of interest, where the recorded seismic signal was not affected by the reservoir zone. Trace-by-trace comparisons provided the best results.

The reflectivity difference shown in Figure 4 was obtained after wavelet processing. A wavelet-shaping filter was applied to the 2000 line to match it to the 1991 line. The most significant differences in the section are seen below the reservoir zone in the area of the production wells. An increase in traveltime through the reservoir zone on the 2000 data was caused by the presence of injected steam in the reservoir. This time delay did not allow the

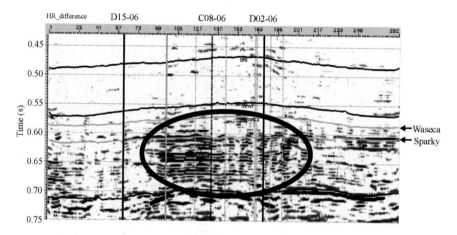

**Figure 4.** Seismic reflectivity–difference section (from Watson et al., 2002). A large difference is seen in the circled area. This area is in the zone of interest at the injector/production well locations. The difference is the result of the time sag associated with injection.

signal of deeper events to cancel when the difference of the sections was computed.

In contrast, the differences are small at the D15-06 location, the well farthest north (shown on the left in Figure 4), where no steam injection or production has occurred. The AI difference, as shown in Figure 5, has some similarities but is not identical to the reflectivity differencing.

An evaluation of reflection times for the Waseca and Sparky reflectors allows for delay-time analysis. In this comparison of time intervals, the H1991 and H2000 geophone array data were used. The use of traveltime intervals eliminates any concern for static differences in the poststack data. The bandwidth differences meant there was greater resolution for picked events on the H2000 stack. There also may be slight tuning differences for the different data vintages because of bandwidth differences. Despite this, the picking of arrival times on a seismic section is robust, and the interpretation of the Waseca-Sparky interval on both versions is shown in Figures 2 and 3.

At each CDP, the ratio of the H2000 to H1991 Waseca interval traveltimes were calculated and plotted, as shown in Figure 6. With the injection of steam and heat in the reservoir in the time between the two surveys, a drop in P-wave velocity is expected. This decrease translates into an increase in the H2000/H1991 isochron ratio.

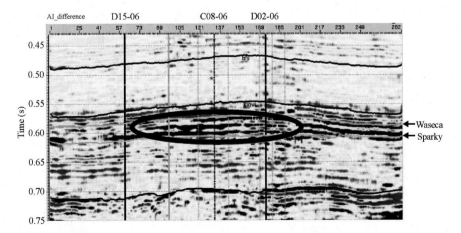

**Figure 5.** Acoustic impedance–difference section (from Watson et al., 2002). The circle marks the area of greatest impedance decrease. The strongest difference is in the Waseca-Sparky interval. The lower impedance does not appear to be constrained to the injector/production area.

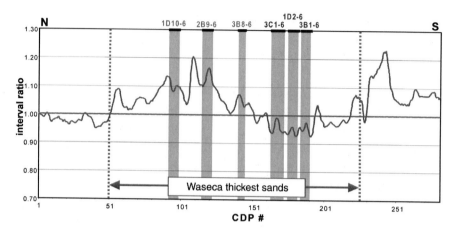

**Figure 6.** H2000/H1991 ratio of Waseca interval traveltimes for P-wave arrivals (from Watson et al., 2002).

The estimated position and width of the steam zone from injection and production data were projected onto Figure 6 from six wellbores near the line (Husky, 2000). The three wells on the left (1D10-6, 2B9-6, and 3B8-6) were drilled during the time period between the two surveys. The ratio rises above unity in this section of the line. Conversely, the ratio drops below unity along the portion of the line where three older producers (3C1-6, 1D2-6, and 3B1-6) were more active in 1991 than in 2000. More heat and steam were present in this portion of the reservoir in 1991 and are responsible for the ratio reversal. These results suggest that the compressional velocity is showing sensitivity to more than just the steam-zone radius around the wellbore. The total area of the heated reservoir also affects $V_P$.

In the studies of multicomponent data at Blackfoot field by Stewart et al. (1996), sand-shale differences were detected using $V_P/V_S$ ratio analysis. In the heavy-oil case at Pikes Peak, the addition of steam into the reservoir decreases $V_P$ and $V_S$. By using core tests for the Waseca samples, the effect of temperature on compressional and shear-wave velocities was investigated. Figure 7 shows that $V_P$ and $V_S$ both decrease with temperature, but $V_P$ decreases at a greater rate. Therefore, it is anticipated that steam injection into a sand unit would cause a decrease in the $V_P/V_S$ ratio.

The $V_P/V_S$ ratio analysis was also interpretation based but involved only the multicomponent data from H2000. No converted-wave data were collected in 1991. The vertical (P-P) and radial (P-S) components were used. For the P-P interpretation, the vertical component of the 3C geophone was used. For the P-S interpretation, the radial component was used. There was no appreciable signal on the transverse component. The interpretation of the radial converted-wave section required a synthetic seismogram that accounted for the wave conversion and the reduced bandwidth on the order of 8–40 Hz.

Figure 3 shows the vertical-component section for H2000, which is predominantly a P-P section. Figure 8 shows the radial-component section, which is predominantly a P-S section. There are two reflection events on each section, which are considered to be depth equivalent, based on the synthetic ties. The Mannville–Lower Mannville intervals are interpreted for the delay-time analysis. The significantly lower bandwidth affects the resolution of picking horizons on the P-S section.

The P-S stacked section also exhibits more noise. Fortunately, at the 1A15-6 tie point, the S/N is relatively high around the zone of interest. Noise cones can be seen on the section to the south (right) of 1A15-6 in the area that coincides with active pump jacks.

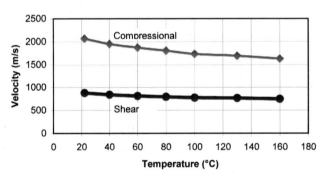

**Figure 7.** Effect of temperature on compressional and shear velocities on a core sample from Pikes Peak D2-6-50-23 (source: Core Laboratories).

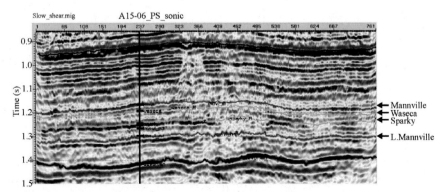

**Figure 8.** H2000 interpreted P-S section (radial component) (from Watson et al., 2002). Note different time scale than in Figure 3.

Using the interval traveltimes, the $V_P/V_S$ ratio is calculated with equation 3:

$$\frac{V_P}{V_S} = \frac{2\Delta t_{PS} - \Delta t_{PP}}{\Delta t_{PP}}, \qquad (3)$$

where $\Delta t_{pp}$ is the traveltime of an interval from the P-P section and $\Delta t_{ps}$ is the interval traveltime from the P-S section. Equation 3 is derived by expressing the thickness of a depth interval in terms of P-wave and S-wave travel.

Figure 9 is a plot of the $V_P/V_S$ ratios for the Mannville–Lower Mannville interval. Noise is still present on this section, but some distinct anomalies can be seen around the wells with the most recent steam injection. In particular, the response at 3B8-6 is a pronounced drop in the $V_P/V_S$ ratio. Steam injection was occurring in this well at the time of the 2000 seismic acquisition. The width of the anomaly fits very well with the predicted steamzone radius. At wells 1D10-6 and 2B9-6, there is a smaller response. It had been 12 and 26 months, respectively, since steam had been injected into those wells.

On a larger scale, the smooth trend line indicates a long-period effect along the length of the line. The low in the middle corresponds very well to the thickest homogeneous Waseca sands, with higher shale content to the north and south. This is similar to the lateral lithology effect that was observed at Blackfoot.

The time-lapse analysis of the two lines shot nine years apart is very encouraging. The reflectivity difference shows the effect of increased traveltime of the seismic signal through the reservoir zone. The impedance difference indicates lower impedance in the reservoir zone.

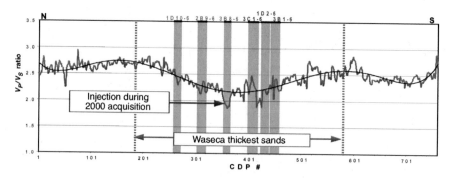

**Figure 9.** $V_P/V_S$ ratio plot of Mannville–Lower Mannville interval (from Watson et al., 2002).

With interpreted seismic sections, the $V_P/V_S$ isochron method provides further insight into the effect of steam injection for heavy-oil reservoirs. In general, the anomalies were located where expected, based on drilling results and injection/production data. A direct steam response can be inferred from the $V_P/V_S$ ratio plots. The time-lapse isochron analysis provided clues about the extent of the heated reservoir. Those results, based on interval interpretations, are very sensitive to tuning and resolution. Bandwidth and phase must be considered carefully during interpretation to ensure that the same depth-equivalent events are being tracked.

Despite those limitations, the methods of detecting steam fronts show general consistency in detecting steam zones. Three of those methods indicate the steam zones through reflection-section differencing, acoustic-impedance differencing, and traveltime differencing. The $V_P/V_S$ ratio method requires multicomponent data but may be a useful detector without the use of a base survey.

## Conclusions

Time-lapse seismology is one of the main tools of reservoir characterization. The main goal is to show differences in the seismic response that are caused only by physical changes in the reservoir. This can be accomplished if the acquisition and processing parameters are as consistent as possible. Thus far, those applications have been applied mainly to mapping of steam zones, but there may be other applications in $CO_2$ and waterflood monitoring as well.

## References

de Buyl, M., 1989, Optimum field development with seismic reflection data: The Leading Edge, **8**, no. 4, 14–20.

Eastwood, J., 1993, Temperature-dependent propagation of P- and S-waves in Cold Lake oil sands: Comparison of theory and experiment: Geophysics, **58**, 863–872.

Eastwood, J., P. Lebel, A. Dilay, and S. Blakeslee, 1994, Seismic monitoring of steam-based recovery of bitumen: The Leading Edge, **13**, no. 4, 242–251.

Hedlin, K., L. Mewhort, and G. F. Margrave, 2002, Delineation of steam floods using seismic attenuation: Canadian Society of Exploration Geophysicists Recorder, **27**, no. 5, 27–30.

Lindseth, R., 1979, Synthetic sonic logs — A process for stratigraphic interpretation: Geophysics, **44**, 3–26.

Lines, L. R., R. Jackson, and J. D. Covey, 1990, Seismic velocity models for heat zones in Athabasca tar sands: Geophysics, **55**, 1108–1111.

Matthews, L., 1992, 3-D seismic monitoring of an in-situ thermal process, *in* R. Sheriff, ed., Reservoir geophysics: SEG, 301–308.

McQuillin, R., M. Bacon, and W. Barclay, 1984, An introduction to seismic interpretation: Gulf Publishing Company.

Nur, A., 1982, Seismic imaging in enhanced recovery: Society of Petroleum Engineers/Department of Energy Transaction 10680.

Pullin, N., L. Matthews, and K. Hirshe, 1987, Techniques applied to obtain very high resolution 3-D seismic imaging at an Athabasca tar sands thermal plot: The Leading Edge, **6**, no. 12, 10–15.

Stewart, R. R., R. Ferguson, S. Miller, E. Gallant, and G. F. Margrave, 1996, The Blackfoot seismic experiments: Broad-band, 3C-3D and 3-D VSP surveys: Canadian Society of Exploration Geophysicists Recorder, **21**, no. 6, 7–10.

Van Hulten, F. F. N., 1984, Petroleum geology of Pikes Peak heavy oil field, Waseca Formation, Lower Cretaceous, Saskatchewan: Canadian Society of Petroleum Geologists, Memoir 9, 441–454.

Watson, I., L. R. Lines, and K. F. Brittle, 2002, Heavy-oil reservoir characterization using elastic wave properties: The Leading Edge, **21**, no. 8, 736–739.

# Chapter 22

# Multicomponent Seismology

Most exploration seismology involves the analysis of P-wave reflections. For waves traveling at near-vertical incidence in a horizontally layered medium, these reflections represent most of the energy recorded by vertical geophones. In marine recording with hydrophone streamers, seismic recordings will consist essentially of P-waves. Conventional seismology has analyzed P-wave reflections with considerable success. Acoustic approximations to the general theories of elastic waves have been fruitful.

Nevertheless, a more complete description of earth's properties is possible through elastic-wave theory, which requires that we consider both P-waves and shear (S-) waves. For this analysis, we use multicomponent recording rather than conventional vertical-component recording.

## P-waves and S-waves

P-waves are compressional longitudinal waves, because the wave's particle displacement is in the same direction as the wave's travel. If the medium is air, these waves are sound waves. P-wave velocity is given by

$$V_p = \sqrt{\frac{k + (4/3)\mu}{\rho}}, \tag{1}$$

where $k$ is the bulk modulus, $\mu$ is the rigidity or shear modulus, and $\rho$ is density.

S-waves are transverse waves, because particle displacement is perpendicular to wave motion. The S-wave velocity (for a homogeneous medium) depends on rigidity (i.e., shear modulus), and it is given by

$$V_s = \sqrt{\frac{\mu}{\rho}}. \tag{2}$$

By comparing equations 1 and 2, we see that P-waves are always faster than S-waves. S-waves will not propagate through fluids, because rigidity equals zero for such materials. Because P-wave displacement is in the direction of wave travel, we record vertically traveling waves with geophones that are sensitive to particle displacement in the vertical direction.

To record S-waves for vertically traveling waves, we would need geophones that are sensitive to particle displacement in the horizontal direction. The horizontal geophones that record displacement in the direction of the seismic line are termed radial phones. Horizontal phones that record displacement perpendicular to the seismic line are termed transverse phones. Three-component receiver units, which have vertical, radial, and transverse orientations, are becoming more common in modern seismic exploration. This allows S-wave information to be used in many applications.

# Why use S-waves?

What are the advantages of recording and analyzing S-wave arrivals as well as P-wave arrivals?

Recall our discussions in Chapter 16 on lithology discrimination and the inversion of P-waves and S-waves. We saw that a knowledge of P-wave velocities generally allows us to distinguish between carbonates and clastics (sandstones and shales). However, within groups of clastic rocks, we need the $V_P/V_S$ ratios to distinguish effectively between sandstones and shales. Therefore, knowledge of S-wave velocities is essential for lithology discrimination.

We also can use S-waves to investigate the orientation and direction of fractures. In exploration seismology, this approach was pioneered by Alford (1986). The generation and recording of S-waves for different orientations allow us to examine the fast and slow modes of S-wave propagation. When a shear wave encounters vertical fractures, as shown in Figure 1, it will split into fast and slow modes of propagation. The fast shear wave has displacement parallel to fractures, and the slow mode has displacement perpendicular to fractures. The density and direction of the fractures will determine the orientation of S-wave modes and the difference in traveltime between fast and slow modes. Hence, this method can help to characterize the fractures within a reservoir and to describe fracture permeability.

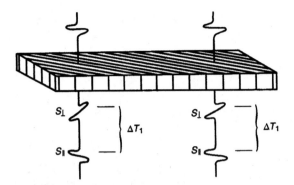

**Figure 1.** Description of S-wave splitting (birefringence) into fast and slow modes on encountering vertical fractures (from Sheriff, 1991).

# Converted-wave exploration

In exploration seismology, S-wave splitting usually is accomplished by generating S-waves using shear-wave sources for different orientations. For this, a shear-wave vibrator generally is used. For a description of vibrators and other S-wave sources, refer to Tatham and McCormack (1991).

The major drawbacks to recording these modes are the cost and the permitting problems that come with use of a shear-wave source. Therefore, most applications use multicomponent recording of conventional sources, such as dynamite or vertical pad vibrators. Such applications use P-S converted waves, the conversion of a downward-propagating P-wave to an upward-propagating S-wave on reflection. A recent paper by Stewart et al. (2002) provides an excellent summary of the state of the art in use of converted waves. Figure 2 shows the geometry of P-wave and P-S reflections.

If we use horizons on both the P-wave and P-S wave seismic sections, we can estimate $V_P/V_S$ ratio by using equation 3 from Chapter 21. This

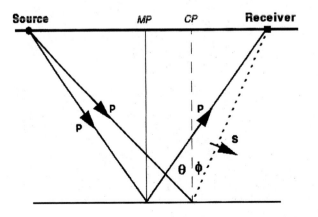

**Figure 2.** Description of P-S converted waves (from Stewart et al., 2002).

would allow for lithology discrimination or the detection of fluids, as in the case of steam flooding.

Many of the applications of P-wave and converted-wave data arise from recent uses of ocean-bottom seismometers. (Because shear waves do not propagate through fluids, marine S-wave exploration requires that we use ocean-bottom receivers.) An example of superior converted-wave data comes from a gas-cloud situation in Tommeliten field of the North Sea, as described by Granli et al. (1999) and by Hoffe et al. (2000).

In Figure 3, we see that gas clouds have a very deleterious effect on the P-wave section. However, the P-S reflection is largely unaffected, because S-waves are not overly sensitive to fluids in rocks. The multicomponent recording allowed an interpretable section to be obtained.

## Conclusions

S-waves and converted P-S waves have been underused in seismic exploration. However, the increased use of multicomponent recording is changing that. Many important emerging applications, such as fracture detection, lithology discrimination, and seeing through gas clouds, are made possible by modern acquisition and processing techniques.

**Figure 3.** Comparison of P-wave stack (top) and P-S stack (bottom) from ocean-bottom recordings for Tommeliten Field (from Granli et al., 1999).

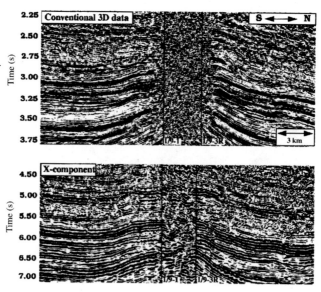

# References

Alford, R. M., 1986, Shear data in the presence of azimuthal anisotropy: 57th Annual International Meeting, SEG, Expanded Abstracts, 476–479.

Granli, J. R., B. Arnsten, A. Sollid, and E. Hilde, 1999, Imaging through gas-filled sediments using marine shear-wave data: Geophysics, **64,** 668–677.

Hoffe, B., L. R. Lines, and P. W. Cary, 2000, Applications of OBC recording: The Leading Edge, **19,** 382–390.

Sheriff, R., 1991, Encyclopedic dictionary of exploration geophysics: SEG.

Stewart, R. R., J. E. Gaiser, R. J. Brown, and D. C. Lawton, 2002, Converted-wave seismic exploration: Methods: Geophysics, **67,** 1348–1363.

Tatham, R. H., and M. D. McCormack, 1991, Multicomponent seismology in petroleum exploration: SEG.

# Chapter 23

# Vertical Seismic Profiles

## Introduction

"A vertical seismic profile (VSP) is a measurement in which a seismic signal generated at the surface of the earth is recorded by geophones secured at various depths to the wall of a drilled well" (Hardage, 1983, p. 1).

There are, in fact, many types of VSP, and the common bond is the borehole. We will look at the acquisition, processing, and interpretation of VSPs from a practical standpoint. VSPs are important because they provide a tie between seismic and borehole images, allow us to create detailed velocity profiles, assist with advanced exploration techniques such as imaging below the drill bit, and allow us to accurately estimate anisotropy parameters.

VSPs have higher resolution than surface seismic recordings, because the seismic waves mostly pass through the attenuating near-surface strata only once, which is not the case with surface-recorded data. A VSP records both downgoing energy and upgoing energy, known as wavefields. Processing of the VSP involves separating the wavefields and identifying the primary events. We interpret the VSP at each processing step to obtain a full understanding of the wavefield.

## Types of VSP

Three main types of VSP are used in seismic exploration. The most common type uses downhole receivers and a surface source (Figure 1) and is what we generally know as a VSP.

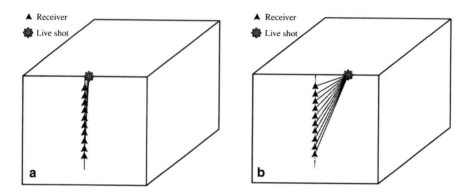

**Figure 1.** VSP (a) zero offset, near offset, or check shot and (b) far offset.

Standard VSP shooting may be categorized into two groups — zero offset and offset VSP. The zero-offset recordings include zero-offset VSP and check-shot surveys in which the source is located at the borehole, as shown in Figure 1a. A check-shot survey differs from a zero-offset VSP with respect to the receiver spacing and the data acquired. The check-shot survey typically is run to check the results of integrating a continuous velocity or sonic log (Sheriff, 1991). The receivers are spread sparsely in the borehole, and only the first breaks are used. Conversely, a zero-offset VSP contains closely spaced receivers and records the entire wavefield.

The offset VSP, as shown in Figure 1b, is a survey in which the source is at a significant distance from the borehole.

A walkaway VSP is a survey that contains different offset VSPs; it also is known as a multioffset VSP. A multiazimuth VSP contains many shots at different azimuths from the borehole. A 3D VSP is usually both multioffset and multiazimuth, with receivers located in a large 3D array around the borehole.

A reverse VSP is acquired by locating the receivers at the surface and the sources within the borehole (Figure 2a). The advantage of a reverse VSP is that a large 3D survey can be conducted quickly by laying out a full surface-receiver array and shooting from within the borehole (Figure 2b). To obtain similar seismic-data coverage from a standard VSP, we would require sources at each of the receiver points and therefore many more times the number of shots. Receivers are more economical than shots and are quicker to deploy in the field.

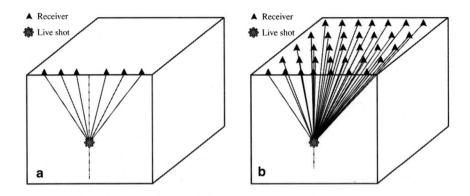

**Figure 2.** Reverse VSP (a) 2D survey and (b) 3D survey with sample raypaths.

A disadvantage of the reverse VSP is possible damage to the borehole, especially if the well is producing. New sources have been shown to have very low impact on the borehole, although damage is still a concern.

Crosswell surveying uses two boreholes, with the receivers in one and the sources in the other (Figure 3). Energy travels directly through the subsurface, and the first-arrival traveltimes can be inverted to obtain a velocity profile of the subsurface. This technique is known as seismic tomography, as described in Chapter 17.

# VSP acquisition

VSP acquisition consists of three main components — a borehole, downhole receivers, and surface source (or, for a reverse VSP, downhole sources and surface receivers). Finding a usable borehole is not trivial. If the well is producing, VSP surveying will shut down production for the length of acquisition. The length of shutdown increases with survey delays, which causes the concern regarding downhole sources. Often, a nearby abandoned well is used.

Sometimes the exploration well must be used so that we can "look ahead of the drill bit" and determine the location in the subsurface to better model subsurface geology. Therefore, it is important to have the survey planned and, ideally, constructed with numerical models to confirm that the parameters are correct for our requirements. So how do we decide on the source and receivers that are required? What offsets do we need? Should this be a 2D or 3D VSP?

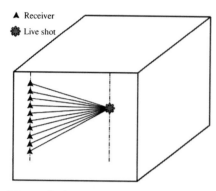

**Figure 3.** Crosswell surveying.

## VSP receivers

Downhole receivers may be one of three types — hydrophone, single component, or three component. Some receiver arrays contain array elements containing both three-component receivers and a hydrophone. Many receivers are located within a single "tool" (e.g., a five-level tool contains five receivers). The receiver spacing within a tool can vary, but our examples use 15-m spacing.

The receiver spacing and the number of levels dictate how the survey will be conducted. If the receiver spacings within the tool are equal to those required for the survey, a conventional approach of pulling the tool up one tool length at a time would suffice (Figure 4a, b). However, other spacings are also possible by interleaving the receiver tool. For example, with a 15-m receiver spacing, we could survey every 7.5 m by shifting the tool by half the receiver spacing for alternate shots, as shown in Figure 4c, d. Contrast this recording with the case of receiver spacings of 15 m, as shown in Figure 4a, b.

The tool is lowered to the bottom of the borehole (or the lowest part of the borehole to be surveyed) and is clamped into position for each shot in that position. When all the shots are completed, the tool is moved up the appropriate distance, reclamped, and held in position for the next series of shots.

If the survey requires many receiver depths to be recorded, the use of a five-level tool can be time-consuming. Paulsson Geophysical Services (P/GSI) has developed tools with 80 levels spanning 10,000 ft (3048 m) and 400 levels spanning 25,000 ft (7630 m). The span of the tool may be sufficient for the desired coverage of the entire survey. If this is the case, the tool may be deployed into the borehole, clamped into place, and held for the duration of the survey. In such situations, there is no need to move the tool or to repeat shots for different levels. If more than 400 levels are required (or if a finer depth spacing needs to be sampled), the tool will need to be moved.

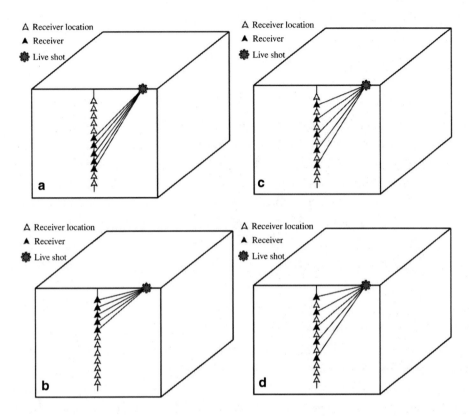

**Figure 4.** When the receiver spacing within the VSP tool equals that required for the VSP survey, the VSP tool is raised up the borehole one tool length at a time. Parts (a) and (b) illustrate two consecutive shots when the tool is raised the length of the tool for each shot. When the receiver spacing within the VSP tool is greater than that required for the VSP survey, the tool is moved up the borehole by interleaving the receiver locations. Parts (c) and (d) illustrate two consecutive shots acquired by interleaving the tool up the borehole.

## VSP sources

Surface sources are much the same as for surface seismic surveys — hammer, thumper, elastic-wave generator (EWG), explosives, and vibroseis. The type of source will depend on the survey being conducted. If we have a full deployment of receivers down the borehole in a single string, any one of the sources mentioned above will suffice as long as it produces the required energy. The local geology and surface conditions would need to be assessed, as in surface seismology.

On the other hand, if we have only a limited number of receivers, e.g., a five-level tool, the source must be repeatable with respect to energy, seismic signature, and location. Therefore, we can pull the tool up the borehole and shoot with the same parameters to obtain comparable shot records. For repeatability, the best option is vibroseis.

Downhole sources include piezoelectric sources, vibrators, and air guns, all of which have been shown to be nondestructive. The piezoelectric and air-gun sources have problems in that they generate tube-wave noise (Paulsson et al., 1998). The downhole vibrators are coupled to the borehole with clamps that minimize the tube waves. The orbital vibrator generates both P- and S-waves simultaneously for all azimuths (Daley and Cox, 2001).

Some studies are investigating the use of the drill bit itself acting as a seismic source so that a continuous reverse VSP can assist in looking ahead of the drill bit (Haldorsen et al., 1995; Malusa et al., 2002).

## VSP processing and interpretation

A basic understanding of VSP processing will assist interpretation of VSPs. Valuable information can be obtained during processing; therefore, it is important to understand the display that is presented. For simplicity, we will consider only zero-offset VSP. Unlike surface seismic surveys, in which a shot record or stack section is presented with distance along the $x$-axis and traveltime along the $z$-axis, the VSP presents receiver depth along the $x$-axis and traveltime along the $z$-axis.

More detailed information describing VSP processing is available in Balch and Lee (1984) and in Hardage (1983).

We use an example from Hinds and Kuzmiski (2000) to illustrate the processing and interpretation steps. The initial shot record presents us with a wealth of information (Figure 5), but note that interference between the upgoing and downgoing energy makes any detailed interpretation difficult. Therefore, the first stage of zero-offset processing is to separate the wavefield into upgoing and downgoing energy (Figure 6).

The first-break curve (or primary downgoing energy) gives a detailed vertical-velocity profile for the borehole (Figure 6b). For every receiver depth (along the *x*-axis), we have a corresponding traveltime (on the *z*-axis). Therefore, we can calculate both average velocities and interval velocities over the survey interval.

Upgoing primary energy gives the depths of the major impedance boundaries. Receivers located at impedance boundaries record both the first-arrival downgoing energy and upgoing energy simultaneously. Although the downgoing wavefield has been removed (as shown in Figure 6c), we can imagine that it is located at the termination of the primary reflection events. Major impedance boundaries are shown with arrows in Figure 6c.

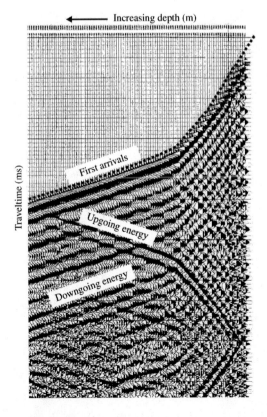

**Figure 5.** Raw VSP shot record with some upgoing and downgoing energy indicated (modified from Hinds and Kuzmiski, 2000).

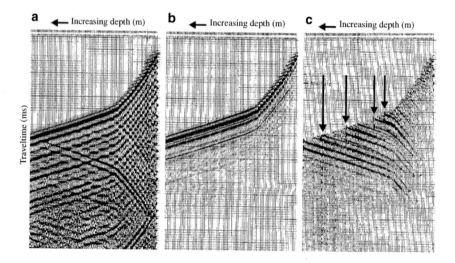

**Figure 6.** (a) Total wavefield recorded, separated into (b) downgoing and (c) upgoing energy. The arrows in (c) indicate some major impedance boundaries (modified from Hinds and Kuzmiski, 2000).

After wavefield separation, multiples are distinguished easily from primary energy. The first-arrival energy is the only downgoing primary energy. All other downgoing wavefields are multiples.

To compare the VSP with surface seismic surveys, we must apply processing techniques such as static time shifting, deconvolution, and muting.

The first concept we must grasp is that of static time shifting. Figure 7 shows downgoing and upgoing raypaths recorded at geophone g. Imagine that the rays are vertical — we present them with a slight inclination (offset) so that they can be seen.

Figure 7a shows downgoing rays recorded at the geophone. Using the nomenclature of Hardage (1983), Ta, Tb, and Tg are the one-way vertical traveltimes to horizon A, horizon B, and the geophone. In viewing Figure 7, we can write expressions for arrival times at the geophone for reflection A, reflection B, and multiple M as

$$\tau_a = 2Ta + \boldsymbol{Tg}$$
$$\tau_b = 2Tb + \boldsymbol{Tg}$$
$$\tau_m = Ta + (Ta - Tb) + (Tg - Tb)$$
$$= 2(Ta - Tb) + \boldsymbol{Tg}.$$

$$(1)$$

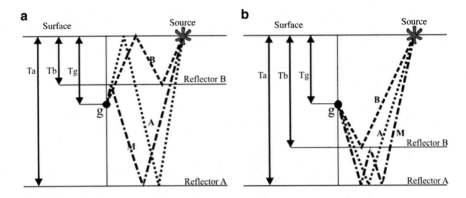

**Figure 7.** Raypaths describing the propagation of (a) downgoing and (b) upgoing energy (modified from Hardage, 1983).

The form of each of these equations is the two-way traveltime of an event at the surface *plus* the one-way traveltime to the geophone. Therefore, if we subtract the one-way traveltime to the geophone (Tg) from each trace, we put each event at the position at which it was last reflected downward. This plot of seismic traces, shifted by subtracting the traveltime to the geophone, is shown in Figure 8 and is referred to as a –TT plot. (Traveltimes are subtracted from the traces).

Conversely, if we examine the upgoing energy in Figure 7b, we obtain the following equations to describe the traveltimes recorded at the geophone:

$$\tau_a = Ta + (Ta - Tg)$$
$$= 2Ta - \boldsymbol{Tg}$$
$$\tau_b = Tb + (Tb - Tg)$$
$$= 2Tb - \boldsymbol{Tg}$$
$$\tau_m = Ta + (Ta - Tg) + 2(Ta - Tb)$$
$$= 2Ta + 2(Ta - Tb) - \boldsymbol{Tg} .$$

$$(2)$$

The form of each of these equations is the two-way traveltime of an event at the surface *minus* the one-way traveltime to the geophone. Therefore, if we add the one-way traveltime to the geophone (the first-break time) to each trace, we obtain the time at which the upgoing events would be recorded at the surface. This addition of traveltime to the traces produces a +TT plot, which is shown in Figure 9.

So how do these two operations of time-shifting the traces help us?

The –TT plot that aligns the downgoing energy (Figure 8) can be used to estimate a deconvolution operator to enhance the VSP image. We will not look at detailed processing techniques such as this. Instead, we will concentrate on the +TT plot, which we can interpret further.

**Figure 8.** (a) Downgoing wavefield and (b) –TT display (modified from Hinds and Kuzmiski, 2000).

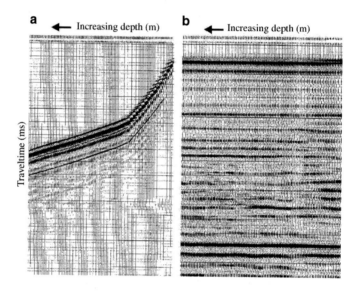

**Figure 9.** (a) Upgoing wavefield and (b) +TT display. The arrows indicate two horizons, before and after transformation (modified from Hinds and Kuzmiski, 2000).

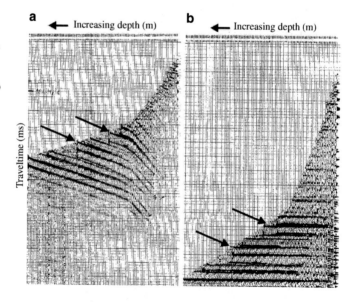

Earlier, we mentioned that multiples mimic primaries but contain a time delay. In the +TT plot (Figure 9), we clearly separate multiples from primaries by noting that they do not intersect the first-arrival energy. After we apply deconvolution to the +TT data (Figure 10a), the multiples become clearer. We observe a small corridor of reflection data that contains only primaries (Figure 10b). This is called a corridor mute, and it is used to form an inside stack containing only primary energy (Figure 10c).

If we stack all the traces prior to muting, we obtain an outside stack that contains both primaries and multiples (Figure 10d). Although it is not shown here, it is also common to build a stack of multiples by subtracting the inside stack from the outside stack.

The outside-corridor stack is useful for determining which energy in our surface seismic is primary and which is from multiples.

All the information gathered during the processing of the zero-offset VSP assists in the integrated interpretation of borehole logs, VSPs, and surface seismic data. For additional readings on VSP, we refer the reader to Hardage (1983) and to Toksöz and Stewart (1984).

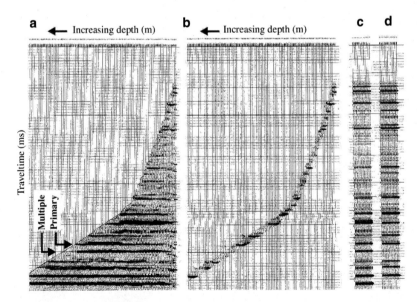

**Figure 10.** (a) +TT display with deconvolution applied, (b) corridor mute, (c) inside corridor stack, and (d) outside corridor stack (modified from Hinds and Kuzmiski, 2000).

# References

Balch, A. H., and M. W. Lee, 1984, Vertical seismic profiling: Technique, applications and case histories: International Human Resources Development Corporation.

Daley, T. M., and D. Cox, 2001, Orbital vibrator seismic source for simultaneous P- and S-wave crosswell acquisition: Geophysics, **66**, 1471–1480.

Haldorsen, J. B. U., D. E. Miller, and J. J. Walsh, 1995, Walk-away VSP using drill noise as a source: Geophysics, **60**, 978–997.

Hardage, B. A., 1983, Vertical seismic profiling — Part A: Principles: Geophysical Press Limited.

Hinds, R. C., and R. D. Kuzmiski, 2000, Exploration VSP: SEG Open File Publications.

Malusa, M., F. Poletta, and F. Miranda, 2002, Prediction ahead of the bit by using drill-bit pilot signals and reverse vertical seismic profiling (RVSP): Geophysics, **67**, 1169–1176.

Paulsson, B. N. P., J. W. Fairborn, and B. N. Fuller, 1998, Imaging of thin beds using advanced borehole seismology: The Leading Edge, **17**, no. 7, 947–953.

Sheriff, R. E., 1991, Encyclopedic dictionary of exploration geophysics: SEG.

Toksöz, M. N., and R. R. Stewart, 1984, Vertical seismic profiling — Part B: Advanced concepts: Geophysical Press Limited.

# Chapter 24

# Cooperative Inversion of Geophysical Data

**Note:** This chapter is excerpted from a paper by Lines et al. which received the SEG award for best paper published in GEOPHYSICS in 1988 (Lines, L. R., A. K. Schultz, and S. Treitel, 1988, Geophysics, **53**, 8–20). It is presented here in a slightly modified version.

## Abstract

Geophysical inversion by iterative modeling involves fitting observations by adjusting model parameters. Both seismic and potential-field model responses can be influenced by adjustment of the parameters of rock properties. The objective of this "cooperative inversion" is to obtain a model which is consistent with all available surface and borehole geophysical data. Although inversion of geophysical data is generally nonunique and ambiguous, we can lessen ambiguities by inverting all available surface and borehole data. This chapter illustrates this concept with a case history in which surface seismic data, sonic logs, surface gravity data, and borehole gravity meter (BHGM) data are modeled adequately by using least-squares inversion and a series of forward-modeling steps.

## Philosophy of cooperative inversion

The traditional analysis of various geophysical data sets has been termed *integrated interpretation*. Such analysis traditionally has involved much processing, modeling, and subjective interpretation to develop a geologic model that is consistent with various surface and borehole geophysical data.

A method that automates some of the above procedures is least-squares inversion. This procedure perturbs an initial set of model parameters to fit the observed data. Least-squares inversion is described in many papers, including those of Jackson (1972), Wiggins (1972), Jupp and Vozoff (1975), and Lines and Treitel (1984). This inversion technique also could be termed *automated iterative modeling*.

In terms of modeling different geophysical data sets, many possible approaches allow inversion of a given data set, and it is doubtful whether one single recipe will prove suitable for all problems. Nevertheless, by jointly inverting various types of geophysical observations, we may lessen the extent of some of the ambiguities inherent in individual data sets. LaFehr (1984) gives a germane discussion of several reasons for this type of coordinated geophysical effort.

In the present study, we define *cooperative inversion* as the estimation of a subsurface model that is consistent with various independent geophysical data sets. In the case discussed in this chapter, these data sets include surface and borehole observations of the seismic and gravity responses. In our definition, cooperative inversion may include either joint inversion or sequential inversion.

## Joint inversion and sequential inversion

The philosophy of joint inversion for seismic and gravity data can be described in the flow diagram of Figure 1. As the diagram outlines, an initial estimate of model densities, velocities, and depths to interfaces is used as input for a seismic model and a gravity-modeling program. The observation set is composed of two parts, seismic (S) and gravity (G). The seismic and gravity model responses, MS and MG, then are computed and compared to the data sets.

In Figure 1, $d_i(x)$ is the thickness of the $i$th layer, which may vary with the horizontal distance $x$; $V_i$ and $p_i$ are the acoustic velocity and density, respectively, of the $i$th layer. An iterative least-squares solution for the model parameters reduces a weighted mean-squared error between observations and the model response. The (preselected) number $W$ (see Figure 1) controls the relative weighting of both responses. Once a satisfactory value of the weighted mean-squared error has been obtained, the final model parameters are the output from the iterative modeling procedure.

Joint inversion involves the simultaneous least-squares inversion of two weighted data sets by perturbation of geologic model parameters. Vozoff and Jupp (1975) used the joint-inversion procedure to model resistivity and magnetotelluric data. Their approach placed weighted values on the two data sets. The weighting of the two different geophysical responses (such as seismic traveltimes and gravity values) in producing a common model is one of the most difficult and important characteristics of this type of inversion. Relative weighting may depend on subjective criteria such as data quality or objective criteria such as sensitivity of the model to changes in parameters. The common parameters in the article by Vozoff and Jupp (1975) were resistivity and cell thickness.

Joint inversion also was applied by Savino et al. (1980) to model teleseismic and gravity data in eastern Washington by varying rock densities

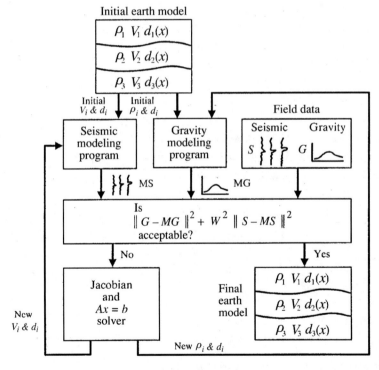

**Figure 1.** The general approach for joint inversion by iterative modeling of seismic and gravity data. $S$ = seismic data, $MS$ = seismic model data, $G$ = gravity data, $MG$ = gravity model data, $\rho_i$ = density for layer $i$, $V_i$ = velocity for layer $i$, and $d_i(x)$ = thickness for layer $i$ (from Lines et al., 1988).

and velocities. To provide the coupling between the density and velocity values in model cells, Savino et al. (1980) used a density-velocity relationship developed by Birch (1961). This relationship is of the form

$$\rho_i = a + bV_i,$$

where $i$ is the cell index and $a$ and $b$ are empirical constants depending on the units of measurement.

The problem of weighting the various data sets can be obviated by use of sequential inversion. In this approach, the inversion for a particular data set provides the input or initial model estimate for the inversion of a second data set. In the sequential inversion investigated here, the geologic boundaries are common to both inversions. The joint and sequential inversion procedures differ in their treatment of the observations. Joint inversion weights the sets of observations and places them into one data vector, whereas sequential inversion treats the sets of observations separately.

One also can categorize cooperative inversions on the basis of parameterization, as described by Golizdra (1980). In Golizdra's description of the integrated seismic and gravity interpretation, the parameterization of inversion can be considered in three types of model categories, which are termed *S, U,* and *M*. In model *S* (for *separate*), no coupling between density and velocity parameters is assumed. Moreover, density-contrast boundaries and velocity-contrast boundaries are assumed to be independent.

Golizdra's model *U* (for *unified*) assumes a coupling relationship between densities and velocities, as well as common boundaries for density and velocity contrasts. This is the type of model used by Savino et al. (1980).

Model type *M* (for *mixed*) strikes a compromise between *S* and *U* and assumes some dependence between density and velocity models. Our proposed sequential inversion uses this semicoupled approach because our density-velocity boundaries are common, but no explicit velocity-density relationship is used.

# Least-squares estimation

In automating our modeling procedures, least-squares algorithms are used to estimate model parameters such as velocity and density. Both unconstrained and constrained algorithms have been used in the estimation procedure. The unconstrained algorithm, described by Lines and Treitel (1984), attempts to minimize the mean-squared error between the data vector and the model-response vector by updating the model parameters.

This minimization procedure requires the solution of the equation

$$A\Delta x = \mathbf{b},\qquad(1)$$

where $A$ is the matrix of partial derivatives of the model response with respect to the model parameters, $\Delta x$ is the parameter change vector, and $\mathbf{b}$ is the discrepancy vector between the data and the model response. If the problem is not linear, several iterations to update the geophysical parameters may be needed. To avoid estimation instabilities, the parameter values can be damped by using the approach of Levenberg (1944) and of Marquardt (1963).

For the constrained optimization problem, an algorithm of Hanson (1982) may be used. This algorithm, bounded and constrained least squares (BOUCLS), solves the least-squares problem and allows constraints to be placed on the model parameters. Equation 1 can be solved subject to bounds on components of the parameter vector (of length $p$) such that

$$\alpha_i < \Delta\chi_i < \beta_i \quad i = 1, 2, \ldots, p\qquad(2)$$

where $\alpha_i$ and $\beta_i$ are lower and upper bounds, respectively, on vector components $\Delta\chi_i$.

The BOUCLS algorithm also allows bounds to be placed on linear combinations of the parameter vector elements. In our problems, we generally use bounded constraints on variables, especially for the case in which parameters are rock densities. Most rock densities should fall between 2.0 and 3.0 g/cm$^3$.

Our strategy generally has been to use the damped least-squares approach until physically unrealistic cases occur. Under such circumstances, the constrained algorithm is used to hold parameter values within reasonable ranges. The damped, unconstrained algorithms generally provide excellent fits to the data, with the possible disadvantage of giving unrealistic parameter estimates. The constrained algorithm has the advantage of parameter control, with the possible disadvantage of too stringent constraints. These may even exclude the current solution.

# The buried fault block — A synthetic test example

To test feasibility of various cooperative inversion strategies, we require forward-modeling solutions for a specific model. The fault-block or wedge model of Figure 2 was chosen for that task because the acoustic seismic response has been given by analytic and finite-difference calculations by Alford et al. (1974). The variables $\rho$ and $V$ are density and velocity, respectively. The gravity response in the far-field limit for such a fault block can be found, for example, in Dobrin (1976).

The gravity inversion for this case is straightforward and is shown in Lines and Treitel (1984). The partial derivatives of the gravity $y$ with respect to density and position coordinates ($cx$, $cz$) are calculated readily, and the least-squares problem then is solved.

The seismic inversion using amplitude data is more difficult because the Jacobian matrix **A** is harder to compute. We can differentiate either the analytical expressions or the finite-difference formulations of Alford et al. (1974) and then solve the least-squares problem. In our case, we used finite-difference, acoustic-wave-equation solutions for the seismic response. The derivatives of these responses were computed using forward differences.

Referring to the wedge problem shown in Figure 2, the seismic source and the receivers were arranged to simulate a marine experiment. The velocity outside the wedge is $V_1$, and inside the wedge it is $V_2 = 0$. Thus, the ratio of the acoustic impedances at an interface is infinite (see Alford et al.,

**Figure 2.** Geometry for joint seismic-gravity synthetic test (from Lines et al., 1988).

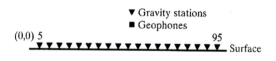

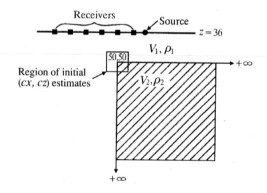

1974). This assumption affects only the amplitudes of the wedge reflections and not the reflection traveltimes. Tests have shown that $V_2 \neq 0$ does not appreciably affect the results of seismic inversion.

The sources and receivers were placed sufficiently deep to avoid surface multiples, although in this case, such multiples are not a major problem. Gravity stations were placed at the surface of this model. These synthetic seismic and gravity data sets were used to locate the wedge corner.

Reasonable wedge locations were obtained from four inversion strategies, including seismic inversion, gravity inversion, sequential inversion, and joint inversion, with two choices of weighting for the two different errors. For one of these joint inversions, we set $W = W_1 = 1.0$. Another joint inversion used a second weight, $W_2$, determined so that the magnitudes of the total seismic error (in milliseconds) and the gravity error (in milligals) were the same. The results are summarized in Table 1. Computation times are given for ten iterations for a less-than-optimum choice of the Marquardt figures. The wedge is on a $100 \times 100$ grid, and the grid points are 22 m (72 ft) apart. The actual position of the wedge is at grid point (50, 50); we used the initial estimate as grid point (46, 46).

After such tests, several preliminary conclusions were drawn. The seismic inversion with finite-difference modeling provides a better solution than the gravity inversion but is much more expensive. The lack of resolution in the gravity inversion is a result partially of the ambiguity between density and anomaly depth.

At this stage, we also decided that ray tracing and traveltime inversion could reduce the seismic inversion cost by at least an order of magnitude. Therefore, traveltime inversion, i.e., kinematic seismic inversion, was used

**Table 1.** Comparisons of inversions for the wedge problem.

| Method | Final wedge position | Computer time for 10 iterations (in CPU seconds) |
|---|---|---|
| Seismic inversion | (51, 50) | 386 |
| Gravity inversion | (49, 48) | 16 |
| Sequential inversion (five iterations gravity, five iterations seismic) | (51, 50) | 227 |
| Joint inversion with $W_1$ | (50, 49) | 429 |
| Joint inversion with $W_2$ | (49, 48) | 429 |

for the real-data examples. Sequential inversion also seemed promising, not only because of the accuracy of the answer and the shorter computation times, but also because the problem of weighting gravity errors versus seismic errors did not need to be addressed. In the joint-inversion procedures, the more accurate answer was provided by the choice $W = 1.0$ (see Figure 1).

Because there were many more seismic samples (600 samples for six traces) than gravity readings (19 gravity station readings), the seismic data were emphasized. Experience with real data also shows the effectiveness of sequential inversion. The only change from our initial sequential-inversion procedure is to implement seismic inversion first on real data to parameterize the problem into layers. This approach will become apparent in our real-data example in the next section.

# Cooperative inversion of a real-data case

After some experience with synthetic and real-data cases, our strategy was to implement sequential inversion. This strategy comes into question later. We illustrate the sequential inversion with a real-data case that included the following data sets:

- CDP stacked seismic sections
- sonic borehole data
- vertical seismic profiles (VSPs)
- borehole gravity meter (BHGM) data
- surface Bouguer gravity maps

We now describe the inversion of these data sets.

# Sequential-inversion strategy

The strategy for first applying the sequential-inversion process to the seismic data sets is outlined in Figure 3. This flow diagram (stage 1 of the inversion process) describes the estimation of a velocity-depth model by use of the available seismic and sonic data. This initial stage requires that the sonic log be "blocked" into a sequence of layers. Given the thicknesses and transit times from the sonic log, we made a first guess for the layer interval velocities by using $V_i = \Delta Z_i / \Delta T_i$, where $\Delta Z_i$ and $\Delta T_i$ are the estimated thicknesses and transit times for the $i$th layer and $V_i$ is the layer velocity. This particular depth-velocity model contains a single interval velocity per layer and insignificant lateral-velocity variations.

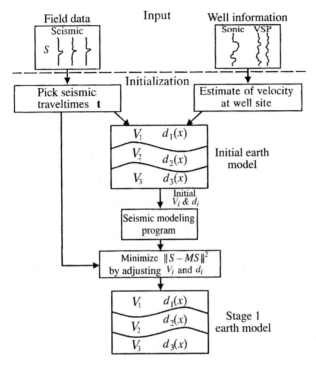

**Figure 3.** Stage 1 of the sequential-inversion problem — the inversion of seismic traveltime data (see Figure 1 for definition of symbols) (from Lines et al., 1988).

The blocked sonic log is shown in Figure 4. The picks on the sonic log agreed reasonably well with the arrivals from a zero-offset VSP. After the initial parameterization of the velocity-depth function at the well site by use of sonic and VSP information, the seismic reflections were interpreted and the traveltimes were picked. The traveltimes provided the set of observations that were to be fitted by modeling. This set provided a much more tractable and manageable database than digitized trace amplitudes. In addition, structural maps of an area will have these times available already.

## Seismic-traveltime fitting

With our set of seismic traveltimes and a velocity-depth function at the well site, we constructed an initial estimate of the subsurface geology. We obtained a first guess for the layer geometry by using a 1D approximation for the depth to the bottom of the $n$th layer $Z_n$:

$$Z_n = \sum_{i=1}^{n} V_i T_i .$$

**Figure 4.** The initial model parameterization in terms of a velocity function from a blocked sonic log (and a corridor-stacked VSP) (from Lines et al., 1988).

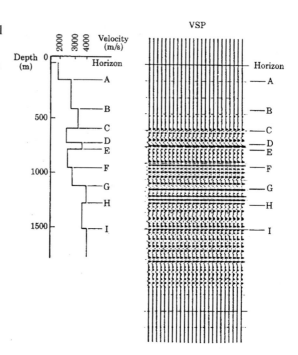

Vertical rays are assumed to be reasonable approximations for the seismic travel paths. For this case, our assumption was justified. The procedure for this "layer-stripping" approach (see below) is shown in the flow diagram of Figure 5, where the velocity $V(x, z)$ depends explicitly on $x$ and $z$.

Our seismic-inversion procedure iteratively uses a normal (or CDP) ray-tracing program along with the least-squares approach of equations 1 and 2. Gjoystdal and Ursin (1981) have described a similar method.

The modeling tools adjusted layer boundaries and velocities to minimize the mean-squared error between the ray-tracing model response and the picked traveltimes. Because there is some ambiguity in depths and velocities, the following procedure was used to decrease the error between the model and the observed traveltimes. The procedure estimates velocities and horizon boundaries from the top layer downward in a layer-stripping approach as follows.

- The model and data traveltimes for a given interface were examined. If there was a relatively constant time shift between the two sets, the velocity was adjusted by least-squares inversion. The velocities also were bounded by use of the sonic log. (The deviation from the sonic log was assumed to be 10% or less.)

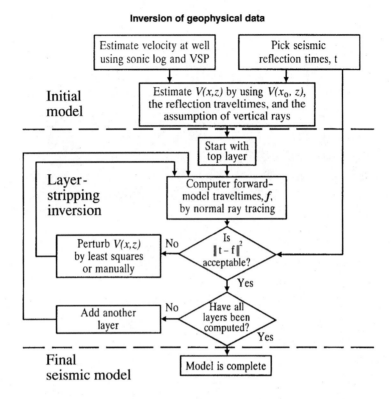

**Figure 5.** The flow diagram for seismic traveltime inversion (from Lines et al., 1988).

- If time discrepancies were not consistently positive or negative, the layer boundary points were adjusted manually to reduce the discrepancy. The results were checked again with an updated ray-tracing model. For the initial seismic model, the average absolute deviation between model and data values was about 2.1 ms.

The application of this procedure to our data is now shown. Figure 6 shows the seismic section and the events, which were picked. The seismic interpretation, which was supplied by Amoco U.K., outlines faulting around a horst feature. The upper pick (A) is highly questionable, and the top boundary was adjusted later to fit the gravity data.

Figure 7 shows the progressive iterative modeling of normal rays (or CDP rays) for some of the layers. The normal rays were traced to compute model-response traveltimes. As pointed out by Hubral (1977), the traveltimes of normal rays should model the time for a CDP stacked section.

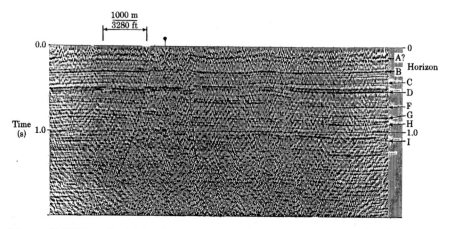

**Figure 6.** CDP stacked seismic time section with annotation of identifiable horizons. Seismic line is 8.1 km, with left edge of the section at x = 0 and well at x = 2.7 km (from Lines et al., 1988).

**Figure 7.** The various ray-tracing models that depict the evolution of a depth model using the procedure outlined in Figure 5 (from Lines et al., 1988).

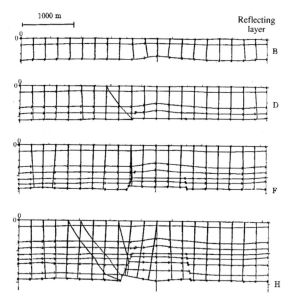

After layer boundaries and velocities had been perturbed to fit traveltimes, the model shown in Figure 8 was produced. A spline fit of the model-response traveltimes of Figure 9a shows an excellent fit to the observed traveltimes of Figure 9b. The average absolute deviation between the data and the model responses for all horizons was 2.04 ms. At this stage, the seismic velocity-depth model was used as input into the gravity-modeling system to compute layer densities and possibly to estimate those layer interfaces that are defined poorly by seismic data.

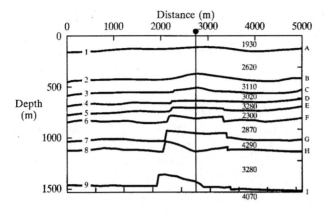

**Figure 8.** The initial seismic model obtained by seismic traveltime inversion. Velocity units are meters per second. The well location is identified at about 2700-m offset (from Lines et al., 1988).

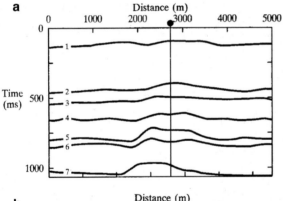

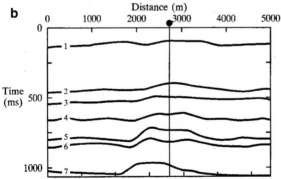

**Figure 9.** (a) Seismic normal ray traveltime (from forward modeling) for the upper layers of the model in Figure 8. (b) Seismic traveltime picks for the upper layers in Figure 8 (from Lines et al., 1988).

## Gravity inversion

After the initial seismic inversion, a 2D velocity-depth model was produced. If we now assume that density and velocity contrasts occur at common boundaries (Golizdra's *M* or *U* model), we can use the geometry from the seismic inversion in a gravity inversion (stage 2 of the sequential-inversion process). This process is outlined in Figure 10. Forward-gravity modeling was implemented with the algorithm of Talwani et al. (1959).

Initial density estimates may come from one of many different sources, such as from well logs or from empirical relations between density and velocity, such as Birch's law (Birch, 1961) or Gardner's equation $\rho = kV^{1/4}$, where $k$ is a proportionality constant (Gardner et al., 1974) and $\rho$ and $V$ are density and velocity, respectively. In the present example, borehole gravimeter (BHGM) data were available. Such valuable information may not always be on hand.

In the first stage, we attempted to use Gardner's equation to infer layer densities from layer velocities (Gardner et al., 1974). This approach corresponds to Golizdra's $U$ model. As shown in Figure 11, the comparison between surface gravity data and the model response is poor across the profile.

Likewise, a comparison between the BHGM response (integrated here between approximate horizons in our coarse model) and the synthetic mod-

**Figure 10.** Flow diagram for the gravity inversion part of sequential inversion (from Lines et al., 1988).

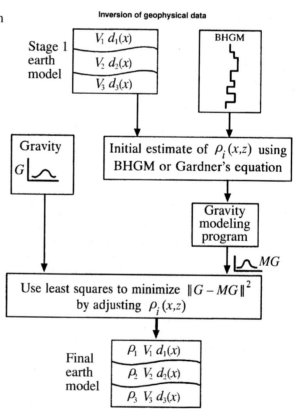

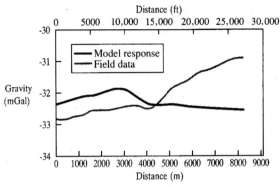

**Figure 11.** Comparison of field data (with increasing trend) to the forward-modeled surface-gravity response using Gardner densities (with flat trend) (from Lines et al., 1988).

el response, as shown in Figure 12, indicates a large discrepancy. This poor correlation possibly is caused by the fact that Gardner's velocity-density relationship generally holds only for clastic sequences. The layer densities were iterated in an attempt to close the gap, but we were unsuccessful. This suggests that our initial density choices were very poor, and we felt compelled to seek a better choice for these starting densities.

Next, densities from compensated formation-density ($\gamma$-$\gamma$) logs were used, but these produced no better fits. The caliper log, which accompanied the compensated formation-density log, showed an unusually high degree of hole rugosity, which may have contributed to the low quality of the log. The usual mud filtrate–invasion problems were also a contributing factor.

Our final choice was the BHGM data set. The density of each layer was equated to the integrated BHGM density between approximate depth horizons (raw field data are shown in Figure 13). The surface gravity and synthetic BHGM responses of the model were calculated. The comparisons are shown in Figures 14 and 15. Figure 14 indicates a dramatic increase in

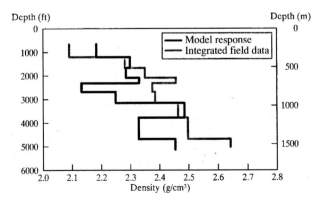

**Figure 12.** Comparison of integrated BHGM field data with the forward-modeled BHGM response using Gardner densities (from Lines et al., 1988).

"goodness of fit" between the surface-gravity field data and the model response in comparison to previous attempts. Naturally, Figure 15 indicates excellent agreement between BHGM (integrated) field data and the model response, because the densities came from the BHGM data set. Any discrepancies between the two plots in Figure 15 result from deviations of the actual layering from the plane-parallel layer case assumed by the Bouguer slab formula for BHGM density determination.

**Figure 13.** BHGM field data (density log) (from Lines et al., 1988).

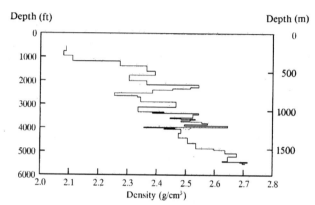

**Figure 14.** Comparison of surface-gravity data with the forward-modeled surface-gravity response using densities obtained from the BHGM log (from Lines et al., 1988).

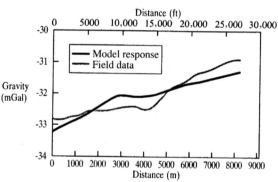

**Figure 15.** Comparison of integrated BHGM field data with the forward-modeled BHGM response using BHGM densities for formation densities (from Lines et al., 1988).

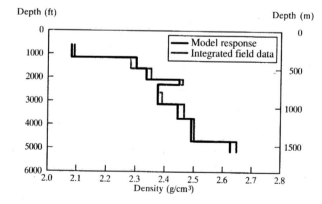

A single unconstrained gravity inversion on the densities then was executed in an attempt to fit the surface profile only. The effects on the model data and the appropriate comparisons are shown in Figures 16 and 17. The fit to the surface-gravity data is much improved, but the lack of constraints on the densities has caused quite a mismatch with the BHGM profile.

Without the comparison between BHGM field data and the BHGM model response, we would have pronounced our inversion a success at that point. However, the extra constraint from an additional geophysical data set restricts the range of acceptable solutions. We thus needed another degree of freedom for our inversion. This degree of freedom required the changes in the model to affect the surface-gravity data but not to affect the BHGM profile drastically.

Our choice was to perturb the boundary between layers 1 and 2 through the inversion process, because this boundary was not defined well on the seismic data. The starting densities were those dictated by the BHGM profile. The results of this inversion, shown in Figures 18 and 19, indicate an acceptable fit between surface and BHGM data.

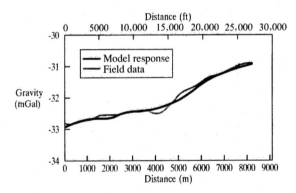

**Figure 16.** Comparison of surface-gravity data with the forward-modeled surface-gravity response after perturbing densities using least-squares inversion (from Lines et al., 1988).

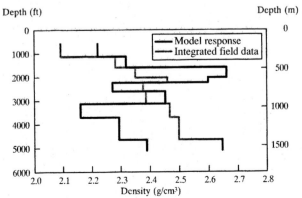

**Figure 17.** Comparison of integrated BHGM field data with the BHGM model response after perturbing densities using bounded and constrained least-squares inversion (from Lines et al., 1988).

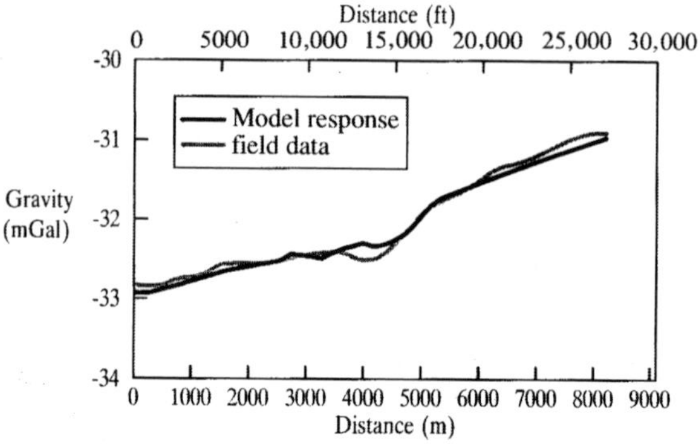

**Figure 18.** Comparison of surface-gravity data with the forward-modeled surface-gravity response after perturbing boundaries between layers (BHGM densities were used for formation densities) (from Lines et al., 1988)

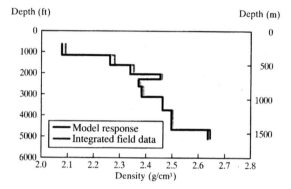

**Figure 19.** Comparison of integrated BHGM field data with the BHGM model response after perturbing boundaries between layers (BHGM densities were used for formation densities) (from Lines et al., 1988)

Some fine-tuning then was done manually on the layer densities, and the model results are shown in Figures 20 and 21. There is a tremendous improvement in the fit between the integrated BHGM data and the BHGM model response, as noted by comparing Figure 21 with Figure 15. The comparison between gravity-data responses at the surface has deteriorated slightly, as noted by comparing Figure 20 with Figure 18. The surface-gravity data are probably valid to $\pm$ 0.1 mGal because of the hazards of positioning and corrections for terrain effects. The BHGM data are probably valid to $\pm$ 0.1 g/cm$^3$. Therefore, the manual adjustment has resulted in a net improvement in fit with both data sets, taking these tolerances into consideration.

Recall that BHGM densities are influenced by two factors. One is the actual bulk density of the various formations. The other is the degree of deviation of formation densities from a Bouguer slab. Actual bulk densities can differ from BHGM densities to compensate for the geometric effects, and that deviation has occurred in this case.

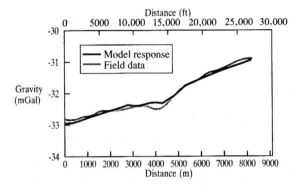

**Figure 20.** Comparison of surface-gravity data with the surface-gravity model response after some manual adjustment (from Lines et al., 1988).

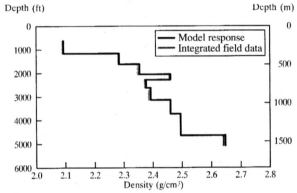

**Figure 21.** Comparison of integrated BHGM field data with the forward-modeled BHGM response after some manual adjustment (from Lines et al., 1988).

# Model consistency and verification

At this point, the effect of model changes on the seismic response must be checked. The model produced by sequential inversion is shown in Figure 22. Because of perturbation of the interface between layers 1 and 2, the seismic-traveltime model did not fit for the lower reflections using the previously estimated velocities for layers 1 and 2. A least-squares inversion was performed to adjust the layer velocities for layers 1 and 2, because these were the layers that did not have well-defined velocities from the sonic well-log data. The least-squares inversion increased the velocity of the upper layer while decreasing the velocity of the second layer.

As indicated in Figure 22, these two upper layers are very close in acoustic impedance (density-velocity product), making this interface almost seismically transparent and possibly accounting for the poorly defined base of Tertiary (horizon A) on the data. Some minor perturbations also were required on the deeper layers and their velocities.

The match of seismic-traveltime responses to the data set for the entire line is shown in Figure 23. The average absolute error for the lines was 8.7 ms, which is acceptable because the error in digitization alone would be about 5.0 ms.

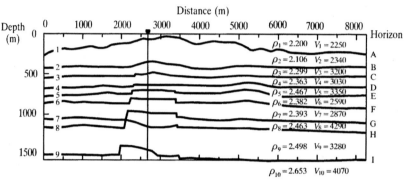

**Figure 22.** Final seismic-velocity depth model. Layer velocities are in meters per second. Layer densities are in grams per cubic centimeter (from Lines et al., 1988).

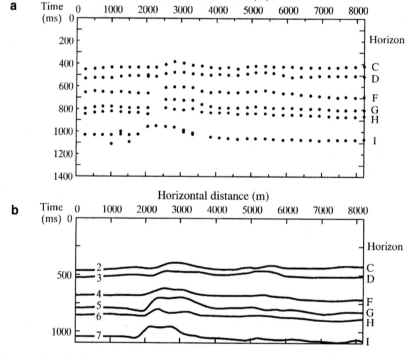

**Figure 23.** (a) Synthetic-traveltime responses for the model compared to (b) the traveltime data picked from the seismic line (from Lines et al., 1988).

In the final analysis, all independent geophysical data sets were reasonably satisfied by the model in Figure 22. The evolution of rock parameters (densities and velocities) and formation depths in the inversion process is shown in Table 2.

This is only one of many possible models that could satisfy the geophysical data, although more data sets narrow the choices. The most uncertain part of the model is the ill-defined interface beneath the upper layer. We have adjusted this layer to fit the gravity data, but it is consistent with the seismic section.

Many other options exist as well. The upper layers may be flat, and the gravity anomaly may be caused by a deep-seated density anomaly. Moreover, 3D effects may be important. We have assumed throughout that the 2D parameterization of our models is reasonable. To verify the strict validity of this assumption, we would need structure maps of all the horizons, and 3D modeling would have to be done.

# Conclusions

In modeling various real-data sets, it became obvious that it would be very difficult (and undesirable) to automate cooperative inversion complete-

**Table 2.** Comparison of density, velocity, and formation depths for initial and final inversion models.

| Layer | Initial density (g/cm³) | Final density (g/cm³) | Initial velocity (m/s) | Final velocity at well | Initial depth estimate at well (m) | Final depth at well (m) |
|-------|-------------------------|-----------------------|------------------------|------------------------|-------------------------------------|-------------------------|
|       | (Gardner density)       |                       |                        |                        |                                     |                         |
| A     | 2.050                   | 2.200                 | 1930                   | 2250                   | 160                                 | 75                      |
| B     | 2.214                   | 2.106                 | 2620                   | 2340                   | 370                                 | 335                     |
| C     | 2.311                   | 2.299                 | 3110                   | 3200                   | 495                                 | 470                     |
| D     | 2.294                   | 2.363                 | 3020                   | 3030                   | 625                                 | 625                     |
| E     | 2.343                   | 2.467                 | 3280                   | 3350                   | 695                                 | 695                     |
| F     | 2.145                   | 2.382                 | 2300                   | 2590                   | 795                                 | 795                     |
| G     | 2.265                   | 2.393                 | 2870                   | 2870                   | 950                                 | 950                     |
| H     | 2.506                   | 2.463                 | 4290                   | 4290                   | 1160                                | 1160                    |
| I     | 2.342                   | 2.498                 | 3280                   | 3280                   | 1420                                | 1420                    |
| J     | 2.472                   | 2.653                 | 4070                   | 4070                   | 1670                                | 1670                    |

ly. Although the methodology for inversion is made easier by many things, such as unconstrained and constrained optimization codes, speedy forward-modeling packages, experienced interpretation, and access to graphics, much of the inversion process still requires the imagination and insight of the geophysicist. The sequential-inversion process appears to be very tractable, because seismic and gravity contributions require no explicit a priori weighting. For this reason, sequential inversion generally is preferred to simultaneous inversion of all data.

In the case presented in this chapter, the coupling between the data sets in the inversion processes arises from the layer geometry. The gravity data enable us to estimate those layer boundaries that are not defined well by seismic data. Whereas the seismic data determine the layer velocities for our model, the surface- and borehole-gravity data sets define layer densities.

In the model fitting of gravity data, a constrained least-squares algorithm proves useful. This algorithm ensures that density estimates honor the borehole gravity data. The algorithm thereby avoids nonphysical density estimates that are often the product of unconstrained inversion. Empirical velocity-density relationships may prove fruitful in some geologic provinces, but in this case, the density estimates were derived from direct physical measurements via BHGM and well logs. This did not require restrictive assumptions on rock type.

The ultimate goal of cooperative inversion is to produce a final earth model that satisfies all the available geophysical data. It is evident that the inversion of several geophysical data sets representing approximately the same portion of the subsurface is less ambiguous than the inversion of any single such data set. Using both seismic and potential-field data makes the choice of geophysical models more restrictive, thus enhancing the interpretation process.

# Acknowledgments

In a project of this scope, there have been many contributors. We thank George Chisholm and Jens Pace of Amoco U.K. for sending us the real-data sets. Seismic- and gravity-modeling assistance was provided by A. Bourgeois, R. Bucker, J. Covey, M. A. Thornton, and K. K. Dill. Finite-difference code and suggestions for modeling the wedge were given by K. Kelly. The constrained optimization code was made available by R. Hanson of Sandia Labs and was transferred to the Amoco system by R. Mercer. We also thank the referees for their excellent suggestions.

# References

Alford, R. M., K. R. Kelly, and D. M. Boore, 1974, Accuracy of finite-difference modeling of the acoustic wave equation: Geophysics, **39**, 834–842.

Birch, F., 1961, Composition of the earth's mantle: Geophysical Journal of the Royal Astronomical Society, **4**, 295–311.

Dobrin, M., 1976, Introduction to Geophysical Prospecting, 3rd ed.: McGraw-Hill Book Co.

Gardner, G. H. F., L. W. Gardner, and A. R. Gregory, 1974, Formation velocity and density — The diagnostic basics for stratigraphic traps: Geophysics, **39**, 770–780.

Gjoystdal, H., and B. Ursin, 1981, Inversion of reflection times in three dimensions: Geophysics, **46**, 972–983.

Golizdra, G. Y., 1980, Statement of the problem of comprehensive interpretation of gravity fields and seismic observations: Izvestiya, Earth Physics, **16**, 535–539.

Hanson, R. J., 1982, Linear least-squares with bounds and linear constraints: Based on Sandia Lab Report SAND82-1517.

Hubral, P., 1977, Time migration — Some ray-theoretical aspects: Geophysical Prospecting, **25**, 738–745.

Jackson, D. D., 1972, Interpretation of inaccurate, insufficient and inconsistent data: Geophysical Journal of the Royal Astronomical Society, **28**, 97–109.

Jupp, D. L. B., and K. Vozoff, 1975, Stable iterative methods for inversion of geophysical data: Geophysical Journal of the Royal Astronomical Society, **42**, 957–976.

LaFehr, T., 1984, Tremors: Why seismic only?: The Leading Edge, **3**, no. 4, 5.

Levenberg, K., 1944, A method for the solution of certain nonlinear problems in least squares: Quarterly of Applied Mathematics, **2**, 164–168.

Lines, L. R., and S. Treitel, 1984, A review of least-squares inversion and its applications to geophysical problems: Geophysical Prospecting, **32**, 159–186.

Marquardt, D. W., 1963, An algorithm for least-squares estimation of nonlinear parameters: Journal of the Society of Industrial and Applied Mathematics, **11**, 431–441.

Savino, J. M., W. L. Rodi, and J. F. Masso, 1980, Simultaneous inversion of multiple geophysical data sets for earth structure: Presented at the 50th Annual International Meeting, SEG.

Talwani, M., J. L. Worzel, and M. Landisman, 1959, Rapid gravity computations for two-dimensional bodies with application to the Mendocino submarine fracture zone: Journal of Geophysical Research, **64**, 49–59.

Vozoff, K., and D. L. B. Jupp, 1975, Joint inversion of geophysical data: Geophysical Journal of the Royal Astronomical Society, **42**, 977–991.

Wiggins, R. A., 1972, The general linear inverse problem: Implication of surface waves and free oscillations for earth structure: Reviews of Geophysics and Space Physics, **10**, 251–285.

# References for general reading

Cooke, D. A., and W. A. Schneider, 1983, Generalized inversion of reflection seismic data: Geophysics, **48**, 665–676.

Crosson, R. S., 1976, Crustal structure modeling of earthquake data, 1: Simultaneous least-squares estimation of hypocenter and velocity parameters: Journal of Geophysical Research, **81**, 3036–3046.

Inman, J. R., 1975, Resistivity inversion with ridge regression: Geophysics, **40**, 798–817.

Rapolla, A., 1981, Some aspects of the interpretation of gravity data, *in* R. Cassinis, ed., The solution of the inverse problem in geophysical interpretation: Plenum Publishing Co.

Vigneresse, J. L., 1977, Linear inverse problem in gravity profile interpretations: Journal of Geophysics, **43**, 193–213.

# Chapter 25

# Geostatistics

One of the main goals of geophysical interpretation is to produce an earth model whose responses fit all the available data (to within some acceptable error tolerance). In Chapter 24, we saw that this was essentially the goal of cooperative inversion.

In this chapter, we will visit briefly with a close cousin of cooperative inversion, geostatistics. Because many textbooks describe geostatistics in more depth than we can here, readers who require more detail are referred to Journel and Huijbregts (1991), Doyen (1988), Isaaks and Srivastava (1989), and Yarus and Chambers (1994).

Geostatistics can provide a set of tools for model prediction and can assign a probability to those predictions. The method typically is used in reservoir characterization to produce model predictions from sets of available geophysical and geologic data. Let us briefly examine a few features of applied geostatistics — forms of prediction known as kriging and cokriging.

## Prediction using a single-variable type

Let us suppose that we have measures of porosity from a group of wells and wish to estimate the porosity at a proposed well location. One approach would be to simply average the well values. This procedure, however, would be susceptible to statistical outliers; that is, estimations would be influenced harmfully by bad data values. A more robust way to handle the situation would be to use spatial prediction with a single physical attribute (other porosity values) or with several physical attributes (such as porosity and seismic amplitudes).

The use of spatial prediction with a single-variable type is known in geostatistics as kriging. Kriging, which is essentially least-squares prediction, was named for D. G. Krige, who applied the method in gold exploration. The method attempts to estimate the value of a variable at a single point by using a weighted combination of surrounding data points. For example, if we hope to estimate the value of a variable, $u$, at a specific location point on a map, $x_0$, we could express the variable in these terms:

$$\hat{u}(x_0) = \sum_{i=1}^{n} w_i u(x_i).$$ (1)

In this case, $x_0$ and $x_i$ refer to vectors defined by Cartesian coordinates on a 2D map. The variable $u$ could refer to some physical quantity, such as porosity or formation depth that we choose to represent in our map.

The model estimate at a specific location is given by weighting values of neighboring points using specific weights, $w_i$ values. These weights are computed so as to minimize the prediction error between the model values and observed values at these points. The least-squares criterion can be used in this minimization. To make the estimate unbiased, we can add the constraint that the sum of all weights be set equal to one (Isaaks and Srivastava, 1989). In a sense, kriging is very similar to least-squares prediction methods used in predictive deconvolution. We can consider kriging as spatial prediction.

## Multivariable prediction

Instead of using a single variable to predict a model, we may wish to use several variable types and increase the information used in the prediction of the model. For example, we may wish to estimate porosity at a specific map location not only by using porosity values from surrounding wells but also by using properties such as 3D seismic-amplitude values that are measured from points throughout the reservoir. In the latter case, we would be performing cokriging (see discussions by Doyen, 1988). Equation 1 would be modified to the following form in which a specific map value, $\hat{u}(x_0)$, is described by values $u(x_i)$ and $v(x_j)$, which are weighted by $w$ and $W$, respectively. In mathematical terms,

$$\hat{u}(x_0) = \sum_{i=1}^{n} w_i u(x_i) + \sum_{j=1}^{m} W_j v(x_j).$$ (2)

In equation 2, the new variables $v(x_j)$ might represent seismic amplitudes.

In some sense, the estimate of a model from cokriging with enough variables could use all the available data measurements obtained from the reservoir. This formulation could achieve the previously stated goals of reservoir characterization and cooperative inversion in which a model is defined by using all available data.

A good example of how cokriging can be used effectively to define a reservoir model was given by Gosse (1999), who showed how seismic data effectively enhanced maps of glauconitic sand-channel models for Little Bow field in southern Alberta.

Figure 1 shows depth estimates of a glauconitic sand channel using kriging. Kriged maps did not provide good images of the channel. Depth errors often exceeded 10 m. Therefore, kriging by itself did not define the channel accurately.

Figure 2 shows cokriged depth estimates that were obtained using both well-log values and 3D seismic data. These values showed errors of 2–5 m

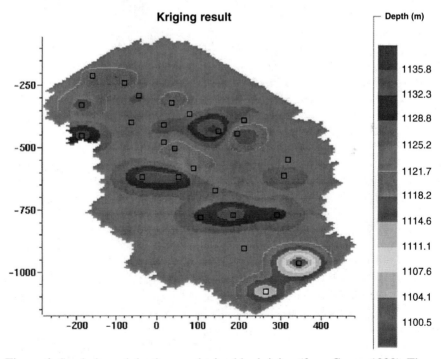

**Figure 1.** Sand-channel depth map obtained by kriging (from Gosse, 1999). The axes represent distances in meters.

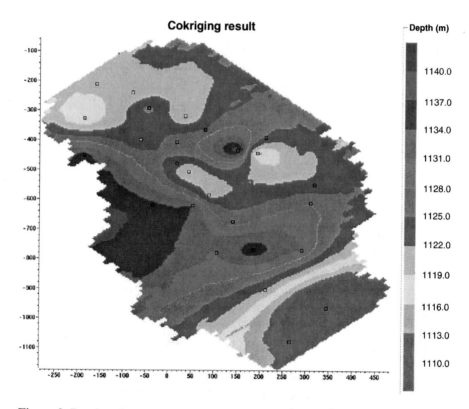

**Figure 2.** Depth estimates for a glauconitic structure using cokriging (from Gosse, 1999). The axes represent distances in meters.

throughout the reservoir. The addition of seismic data significantly improved the definition of the channel.

## Conclusions

Gosse's geostatistical results agree with the cooperative inversion study in Chapter 24. As we add more geophysical information to our model estimation, we improve the accuracy of our earth-model results. Both cooperative inversion and geostatistics provide the basic tools for reservoir characterization.

# References

Doyen, P. M., 1988, Porosity from seismic data: A geostatistical approach: Geophysics, **50**, 1263–1275.

Gosse, C., 1999, Geostatistical analysis of Little Bow field, Alberta, Canada: M.S. thesis, University of Calgary.

Isaaks, E. H., and R. M. Srivastava, 1989, An introduction to applied geostatistics: Oxford University Press.

Journel, A. G., and C. J. Huijbregts, 1991, Mining geostatistics: Academic Press Inc.

Yarus, J. M., and R. Chambers, 1994, Stochastic modeling and geostatistics: Principles, methods and case studies: AAPG Computer Applications in Geology No. 3.

# Chapter 26

# The Art and Science of Contouring

Most significant petroleum exploration plays are accompanied by a series of contour maps representing the subsurface geology. Today, seismic maps are generally the foundation of a well recommendation or seismic-survey proposal. The following uses the discussions of Tucker (1988) and of Gadallah (1994) in giving a brief summary of contour mapping.

It is essential that the contour map convey to management the essential message. In the words of Tucker (1988, p. 741), "Here's the oil deposit!" We can interchange *gas* or *mineral deposit* for *oil* in that statement, depending on the play.

Most of us are familiar with topographic maps that contain lines of constant elevation, known as elevation contours. The contours of a subsurface structural map for a given formation display the topography as it would appear if rocks above the formation were stripped away. Such contour maps display the dip of a formation, as well as any folding or faulting of the structure.

Maps can be constructed to display the depth of a formation from some datum level. The data for seismic contour maps are obtained by picking times or depths from seismic sections. Because the picks from the seismic section are the essential data for our map, they should be made with considerable care. Generally, the initial picks are made by using ties to well logs or formation tops. The times and depths are then picked through the seismic-data volume by correlating reflected arrivals from trace to trace.

It is worthwhile to identify the quality of seismic picks as good, fair, or poor. Quality is often a function of signal-to-noise ratio and will influence how well one honors certain points.

If there is a well-defined reference horizon on a seismic section, we may map the differences in time (or depth) from the reference horizon to the horizon of interest. If depths are picked, an isopach map is created; if times are picked, an isochron map is created. In other words, an isopach map shows variation of formation thicknesses. An isochron map shows variation of seismic-reflection times between horizons.

# Contouring — An interpretive process

The art and science of contouring are more than mere mechanical procedures. Contouring should be done with a particular geologic concept in mind. The mechanics can be governed by a few simple rules, included in the discussions of Gadallah (1994):

- Isolate the highest and lowest values in the collection of data points.
- A contour line should pass between points whose numerical values are higher and lower than the value of the contour line.
- A contour never crosses over itself or another contour except in the case of reverse faults or overturned folds.
- A contour cannot merge with contours of different values.
- For valleys or streams, the contours always point upstream.
- Use dashed lines when control points are lacking.
- Keep the map as simple as possible, while still honoring points.

# Caveats about machine contouring

Contouring is usually difficult and time-consuming, but it is crucial to the exploration play. For this reason, computer contouring often is used, but one should interpret computer-contoured maps with a grain of salt. Although such maps can be effective for showing structures, they rarely do a good job of delineating fault structures.

The example in Figure 1 shows some of the problems that can arise from machine contouring. The hand-contoured version describes the folds and the wrench fault and looks more geologically feasible. The shapes of the contours are not square but have the flow of folded beds.

Computer programs can contour data effectively for simple geologic structures, but most computer algorithms do not include geologic concepts effectively and do not handle fault discontinuities well. Computer algorithms also may be overly sensitive to noisy picks, giving high-frequency contour wiggles that are not geologically realistic.

a

b

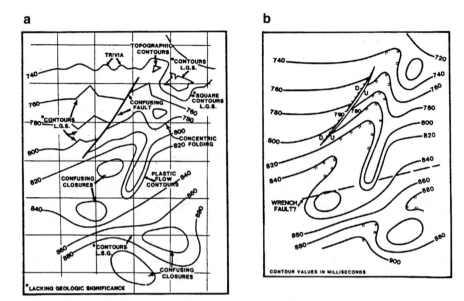

**Figure 1.** (a) Computer contouring and some of its inherent problems, contrasted with (b) a hand-contoured map (from Tucker, 1988).

It is probably best to avoid a total reliance on computer-contoured maps for complicated structural plays — especially when geologic concepts are not included in the map inherently.

Faults pose the biggest challenge for computer mapping. For a good discussion of effectively contouring faults by computer, the reader is referred to Tucker (1988).

# Pitfalls in contouring

We examined above some of the basic rules for computer contouring. Tucker (1988) points out several pitfalls. Some of the most important are:

- assuming that contour values are "gospel" (i.e., absolutely correct); seismic picks will have errors and should not be taken too literally, so as to make each contour bend around the value

- using a contour interval that is too small, thereby creating a map that emphasizes noise rather than main geologic trends

- contouring for an inappropriate geological style; background knowledge of an area's geology will help avoid this

- mapping only one horizon
- herringbone pattern contouring, usually caused by mislocation of a seismic line
- producing square contours
- using too many faults, resulting in a "cracked-glass" appearance

## Maps for prospect generation

In terms of well recommendations to management, experience is the best teacher. Too often, a good prospect can be declined on the basis of an unclear or poorly contoured map. The map should clearly outline the geologic structural style and the reasons for a well location. The map should integrate the concepts of land position, geology, geophysics, and reservoir engineering.

Tucker (1988) points this out clearly with two maps, shown in Figure 2. One has too many faults that are not defined clearly, giving a "cracked-

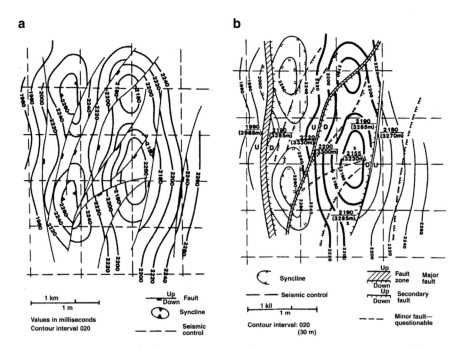

**Figure 2.** Two maps contrast (a) the "cracked-glass" appearance of too many faults with (b) the desired prospect map (from Tucker, 1988).

glass" appearance. The second shows some well-defined faults in which the upthrown and downthrown sides are defined clearly, as are anticlines, synclines, horizon depths, and questionable faults. Furthermore, the second map clearly defines closure, fault displacement, seismic control, and depth of the prospect.

## Conclusions

The seismic-contour map is a very important part of the exploration play. It is essential for the generation of a petroleum prospect. Therefore, the art and science of contouring are key skills for any geophysical interpreter.

Although the simple rules for contouring given in this chapter will prove helpful, there is no substitute for the actual experience of contouring maps, combined with ongoing discussions among geologists and geophysicists.

## References

Gadallah, M. R., 1994, Reservoir seismology: PennWell Books.

Tucker, P. M., 1988, Seismic contouring: A unique skill: Geophysics, **53**, 741–749.

# Chapter 27

# Conclusions

This book on interpretation covers several fundamental geophysical principles and is intended to serve as an introduction to the topic. It is certainly not the final word on the subject.

We hope the reader realizes, by this point, that interpretation is an integrative process. Effective interpretation will use all available data and will produce a map or model that is consistent with all observations. Tools such as cooperative inversion or geostatistics indicate that interpretation is enhanced by the use of several reliable data sets. This is definitely true of reservoir characterization, which requires data of different wavelengths to provide a complete picture of the reservoir volume.

Much has changed in geophysical interpretation in past decades. Although reflection seismology remains the workhorse of geophysical exploration, 3D seismic surveys are now the standard reservoir characterization tool and are used routinely. The seismic-interpretive workstation has allowed vast amounts of 3D data to be analyzed effectively in a tractable manner. Visualization tools have seen great advances in the last decade. One of the greatest advances in fault detection has come about through development of coherency technologies.

Pitfalls in seismic interpretation will always be with us. However, many of the pitfalls of the last decade have been obviated by use of modern exploration methods.

One of the most common pitfalls in past years has arisen from misinterpretation of seismic time-section anomalies as depth anomalies. This can be avoided by effective use of depth migration. However, this solution assumes that we have obtained accurate velocity estimates — the velocity-estimation problem continues to be important.

The pitfall of "out-of-plane" reflections can be solved by use of 3D seismic exploration. Modern multiple-suppression methods should help to eliminate the pitfall of interpreting multiples as primaries. Although we have seen significant progress in suppression of multiples, we have only recently become aware of pitfalls related to anisotropy. For instance, isotropic imaging of anisotropic media will cause errors in location of images. The cure for this problem requires anisotropic depth imaging with accurate velocity and anisotropy estimates for dipping sediments.

Interpretation is a combination of both art and science. It does not lend itself to automation but requires the use of skilled human intervention and insight, combined with advanced computer hardware and software. Finally, interpretation is the most important process for development of petroleum and minerals, and it will continue to advance to match the world's thirst for natural resources.

# Index